4/1/93

Introduction to Lattices and Order

D0165721

Introduction to Lattices and Order

B.A. Davey
Reader in Mathematics,
La Trobe University

H.A. Priestley
Lecturer in Mathematics in the
University of Oxford

CAMBRIDGE
UNIVERSITY PRESS

Published by the Press Syndicate of the University of Cambridge
The Pitt Building, Trumpington Street, Cambridge CB2 1RP
40 West 20th Street, New York, NY 10011–4211, USA
10 Stamford Road, Oakleigh, Victoria 3166, Australia

First published 1990
Reprinted 1991, 1992

Printed in Great Britain at the University Press, Cambridge

British Library cataloguing in publication data available

Library of Congress cataloguing in publication data available

ISBN 0 521 36584 8 hardback
ISBN 0 521 36766 2 paperback

Contents

Preface

This is the first textbook devoted to ordered sets and lattices and to their contemporary applications. It acknowledges the increasingly major role order theory is playing on the mathematical stage and is aimed at students of mathematics and at professionals in adjacent areas, including logic, discrete mathematics and computer science.

Lattice theory has been taught to undergraduates at La Trobe University since 1975, and more recently at Oxford University. The notes for these courses were our starting point. The core of the book— Chapters 1, 2 and 5 to 8—provides a basic introduction to ordered sets, lattices and Boolean algebras and is buttressed by exercises which have been classroom-tested over many years. In a proselytizing article, *Order: a theory with a view* [31], Ivan Rival discusses the modern role of order. The pictorial philosophy he advocates is strongly evident in our approach: diagrams and diagrammatic arguments are stressed in both the text and the exercises.

Prerequisites are minimal. A reader who has taken a course in linear algebra, group theory or discrete mathematics should have sufficient background knowledge and be familiar with our vocabulary and with those symbols not listed in the notation index. To keep the treatment as elementary as possible, we have denied ourselves the formalism of category theory and of universal algebra. However we have prepared the ground carefully for those who will progress to texts on general lattice theory or universal algebra and we have included, at the ends of Chapters 2 and 5 and in Chapters 3, 9 and 10, some material suitable for honours students or those beginning graduate work. Inevitably, there was not space for all the topics we should have liked to cover; hints of resisted temptations will be apparent in a few of the exercises. Within lattice theory we have placed the emphasis on distributive lattices. We thereby complement more advanced texts, in which modular and general lattices are already well treated. The study of finite distributive lattices (undertaken in Chapter 8) combines algebraic, order-theoretic and graph-theoretic ideas to provide results which are linked to the ordered set constructions presented in Chapter 1, are easily accessible to undergraduates and are complete in themselves. Our colleagues will doubtless not be surprised that we have also included the extension of the representation theory to the infinite case. To coax those wary of topology, this introduction to duality is accompanied by a self-contained primer containing the small number of topological results which we need.

Order has recently appeared, sometimes a little coyly, in many computational models. The thorough treatment of ordered sets in Chapter 1 (with examples foreshadowing applications in computation) and of intersection structures in Chapter 2 provides a firm foundation on which to build the theory of CPOs and domains. Chapter 3 studies these structures and relates them to Scott's information systems. Our account is necessarily brief. Collateral reading of specialized texts, in which the computer science applications are fully developed, may assist those meeting domain theory for the first time. Chapter 4 deals with fixpoint theory (and also discusses the order-theoretic roots of Zorn's lemma). Thus Chapters 1–4 serve as an introduction to order theory for computer scientists, and for mathematicians seeking to enter their world. In Chapter 11 we look outwards in a different direction and present the rudiments of formal concept analysis. This new field has already made an impact on lattice theory and has much to offer to social scientists concerned with data analysis. We acknowledge our debt to the authors of many unpublished notes and manuscripts on computer science and on concept analysis. In particular, course notes by Dana Scott, Samson Abramsky and Bill Roscoe enticed us into previously unfamiliar territory and Jeff Sanders' notes for the hardware course taught to Mathematics and Computation undergraduates in Oxford influenced our treatment of Boolean algebras.

The technological developments of the 1980s have made our collaboration possible. Our respective computers have faithfully worked many nocturnal hours of overtime. Electronic mail has enabled us to communicate almost daily and, in conjunction with TₑX, to confer easily on fine points of presentation in a way that would have been impossible with conventional ("snail") mail. TₑX has also allowed us to control the final shape of the text and has given the second author in particular innumerable hours of fun and frustration.

Many people deserve our thanks. We are grateful to David Tranah and the staff of Cambridge University Press for their patient assistance and support and to Dorothy Berridge for her help in typing TₑX files. Generations of students have provided valuable consumer feedback and Oxford undergraduates Mark Joshi, Graham Pollitt and Andy Sanderson earn a special mention for their proof-reading. Thanks are due to the colleagues we have pestered to read the book in draft: in particular to Michael Albert, Ralph McKenzie and J.B. Nation. We must also thank Rudolf Wille and Bernhard Ganter for their advice on concept analysis. Notwithstanding electronic communication, we greatly benefited from the opportunity to spend a month discussing the book face to face. The second author gratefully acknowledges the financial assistance of La Trobe University and the hospitality of its mathematics department. Finally, a very big thank you for their support and forebearance goes to the Davey family: wife Helen and children Evan, Owen and Caitlin.

B.A.D. and H.A.P.

September 1989

1

Ordered Sets

Order, order, order—it permeates mathematics, and everday life, to such an extent that we take it for granted. It appears in many guises: first, second, third, ... ; bigger versus smaller; better versus worse. Notions of progression, precedence and preference may all be brought under its umbrella. Our first task is to crystallize these imprecise ideas and to formalize the relationship of 'less-than-or-equal-to'. Besides presenting examples and basic properties of ordered sets, this chapter also introduces the diagrams which make order theory such a pictorial subject and give it much of its character.

Ordered sets

What exactly do we mean by order? More mathematically, what do we mean by an ordered set?

1.1 Order. Each of the following miscellany of statements has something to do with order.

(a) $0 < 1$ and $1 < 10^{23}$.

(b) Two first cousins have a common grandfather.

(c) 22/7 is a worse approximation to π than 3.141592654.

(d) The planets in order of increasing distance from the sun are Mercury, Venus, Earth, Mars, Jupiter, Saturn, Uranus, Neptune, Pluto.

(e) Neither of the sets $\{1, 2, 4\}$ and $\{2, 3, 5\}$ is a subset of the other, but $\{1, 2, 3, 4, 5\}$ contains both.

(f) Given any two distinct real numbers a and b, either a is greater than b or b is greater than a.

Order is not a property intrinsic to a single object. It concerns comparison between pairs of objects: 0 is smaller than 1; Mars is further from the sun than Earth; a seraphim ranks above an angel, etc. In mathematical terms, an ordering is a binary relation on a set of objects. In our examples, the relation may be taken to be 'less than' on $\mathbb{N}$ in (a), 'is a descendant of' on the set of all human beings in (b) and $\subseteq$ on the subsets of $\{1, 2, 3, 4, 5\}$ (or of $\mathbb{N}$) in (e).

What distinguishes an order relation from some other kind of relation? Firstly, ordering is transitive. From the facts that $0 < 1$ and $1 < 10^{23}$ we can deduce that $0 < 10^{23}$. Mars is nearer the sun than Saturn and Saturn is nearer than Neptune, so Mars is nearer than Neptune.

Secondly, order is antisymmetric: 5 is bigger than 3 but 3 is not bigger than 5. It is on these two properties—transitivity and antisymmetry—that the theory of order rests.

Order relations are of two types: strict and non-strict. Outside mathematics, the strict notion is more common. The statement 'Charles is taller than Bruce' is generally taken to mean 'Charles is strictly taller than Bruce', with the possibility that Charles is the same height as Bruce not included. Mathematicians usually allow equality and write, for instance, $3 \leqslant 3$ and $3 \leqslant 22/7$. We shall deal mainly with non-strict order relations.

Finally a comment about comparability. Statement (f) asserts that, for the ordering $<$ on the real numbers, any two distinct elements can be compared. This property is possessed by many familiar orderings, but it is not universal. For example, there certainly exist human beings A and B such that A is not a descendant of B and B is not a descendant of A. Non-comparability also arises in (e).

1.2 Definitions.   Let P be a set. An **order** (or **partial order**) on P is a binary relation $\leqslant$ on P such that, for all $x, y, z \in P$,

 (i) $x \leqslant x$,

 (ii) $x \leqslant y$ and $y \leqslant x$ imply $x = y$,

 (iii) $x \leqslant y$ and $y \leqslant z$ imply $x \leqslant z$.

These conditions are referred to, respectively, as **reflexivity**, **anti-symmetry** and **transitivity**. A set P equipped with an order relation $\leqslant$ is said to be an **ordered set** (or **partially ordered set**). Some authors use the shorthand **poset**. Usually we shall be a little slovenly and say simply 'P is an ordered set'. Where it is necessary to specify the order relation overtly we write $\langle P; \leqslant \rangle$.

An order relation $\leqslant$ on P gives rise to a relation $<$ of **strict inequality**: $x < y$ in P if and only if $x \leqslant y$ and $x \neq y$. It is possible to re-state conditions (i)–(iii) above in terms of $<$, and so to regard $<$ rather than $\leqslant$ as the fundamental relation; see Exercise 1.1.

Other notation associated with $\leqslant$ is predictable. We use $x \leqslant y$ and $y \geqslant x$ interchangeably, and write $x \nleqslant y$ to mean '$x \leqslant y$ is false', and so on. Less familiar is the symbol $\parallel$ used to denote non-comparability: we write $x \parallel y$ if $x \nleqslant y$ and $y \nleqslant x$.

We later deal systematically with the construction of new ordered sets from existing ones. However there is one such construction which it is convenient to have available immediately. Let P be an ordered set and let Q be a subset of P. Then Q inherits an order relation from P;

given $x, y \in Q$, $x \leqslant y$ in Q if and only if $x \leqslant y$ in P. We say in these circumstances that Q has the order **induced from** P.

1.3 Chains and antichains. Let P be an ordered set. Then P is a **chain** if, for all $x, y \in P$, either $x \leqslant y$ or $y \leqslant x$ (that is, if any two elements of P are comparable). Alternative names for a chain are **linearly ordered set** and **totally ordered set**. At the opposite extreme from a chain is an antichain. The ordered set P is an **antichain** if $x \leqslant y$ in P only if $x = y$. Clearly, with the induced order, any subset of a chain (an antichain) is a chain (an antichain).

Let P be the n-element set $\{0, 1, \dots, n-1\}$. We write **n** to denote the chain obtained by giving P the order in which $0 < 1 < \cdots < n-1$ and $\overline{\mathbf{n}}$ for P regarded as an antichain. Any set S may be converted into an ordered set $\overline{S}$ by giving S the antichain order.

Examples from mathematics, computer science and social science

We hinted in 1.1 at a variety of situations in which order structure is present. In 1.2 we developed the vocabulary for treating these examples systematically. This section is a catalogue of ordered sets, drawn from mathematics, computer science and the social sciences.

1.4 Number systems. The set of real numbers, $\mathbb{R}$, forms a chain in its usual order. Each of $\mathbb{N}$ (the natural numbers $\{1, 2, 3, \dots\}$), $\mathbb{Z}$ (the integers) and $\mathbb{Q}$ (the rational numbers) also has a natural order making it a chain. In each case this order relation is compatible with the arithmetic structure in the sense that the sum and product of two elements strictly greater than zero is also greater than zero. On the other hand, the complex numbers, $\mathbb{C}$, carry no order relation with this compatibility property; see Exercise 1.24.

We denote the set $\mathbb{N} \cup \{0\}$ $(= \{0, 1, 2, \dots\})$ by $\mathbb{N}_0$. Endowed with the order in which $0 < 1 < 2 < \dots$, the set $\mathbb{N}_0$ becomes the chain known in set theory as ω. A different order on $\mathbb{N}_0$ is defined as follows. Write $m \preccurlyeq n$ if and only if there exists $k \in \mathbb{N}_0$ such that $km = n$ (that is, m divides n). Then $\preccurlyeq$ is an order relation. Of course, $\langle \mathbb{N}_0; \preccurlyeq \rangle$ is not a chain. Yet another order on $\mathbb{N}_0$ is introduced in 1.24 for use in Chapters 3 and 4.

1.5 Families of sets. Let X be any set. The powerset $\wp(X)$, consisting of all subsets of X, is ordered by set inclusion: for $A, B \in \wp(X)$, we define $A \leqslant B$ if and only if $A \subseteq B$.

Any subset of $\wp(X)$ inherits the inclusion order. Such a family of sets might be specified set-theoretically. For example, it might consist of all finite subsets of an infinite set X. More commonly, families of sets arise where X carries some additional structure. For instance, X might have an algebraic structure—it might be a group, a vector space, or a ring. Each of the following is an ordered set under inclusion:

the set of all subgroups of a group G (denoted $\text{Sub}\,G$), and the set of all normal subgroups of G (denoted $\mathcal{N}\text{-Sub}\,G$);

the set of all subspaces of a vector space V;

the set of all subrings of a ring, R, and the set of all ideals of R.

Families of sets also occur in other mathematical contexts. For example, let $(X; \mathcal{T})$ be a topological space. We may consider the families of open, closed, and clopen (meaning simultaneously closed and open) subsets of X as ordered sets under inclusion. Finally we note a more inbred member in this class of ordered sets which is of fundamental importance later. This is the family $\mathcal{O}(P)$ of down-sets of an ordered set P; it is defined in 1.19.

1.6 Information orderings. The statement that some computed quantity r equals 1.35 correct to 2 decimal places may be re-expressed as the assertion that r lies in a particular closed interval in **R**. We may accordingly treat the collection of all intervals $[\underline{x}, \overline{x}]$ (where $-\infty \leqslant \underline{x} \leqslant \overline{x} \leqslant \infty$) as defining a set P of approximations to the real numbers, with the intervals for which $\underline{x} = \overline{x}$ corresponding to exact values. The set P carries a very natural order: for $x = [\underline{x}, \overline{x}]$ and $y = [\underline{y}, \overline{y}]$ define $x \leqslant y$ if and only if $\underline{x} \leqslant \underline{y}$ and $\overline{y} \leqslant \overline{x}$. Then $x \leqslant y$ means that y represents (or contains) at least as much information as x.

We next consider strings. Let Σ^* be the set of all finite binary strings, that is, all finite sequences of zeros and ones; the empty string is included. Adding the infinite sequences, we get the set of all finite or infinite sequences, which we denote by Σ^{**}. We order Σ^{**} by putting $u \leqslant v$ if and only if u is a finite initial substring (the technical term is **prefix**) of v. Thus, for example, $0100 < 010011$, $010\|100$ and $10101 < 101010\ldots$ (the infinite string of alternating zeros and ones). Strings may be thought of as information encoded in binary form: the longer the string the greater the information content. Further, given any string v, we may think of elements u with $u < v$ as providing approximations to v. In particular, any infinite string is, in a sense we shall later need to make precise, the limit of its finite initial substrings. Obviously this example can be generalized. We may consider strings whose elements are drawn from an arbitrary alphabet of symbols, and order it in the same way as above.

Our final class of examples in this group concerns partial maps.

Let X be a set and consider a map $f\colon X \to X$. Then f may be regarded as a recipe which assigns a member $f(x)$ of X to each member x of X. Alternatively, and equivalently, f is determined by its graph, $\operatorname{graph} f := \{\,(x, f(x)) \mid x \in X\,\}$, a subset of $X \times X$. If the values of f are given on some subset S of X, we have partial information towards determining f. Formally we define a **partial map** on X to be a map $\sigma\colon S \to X$, where $\operatorname{dom}\sigma$, the domain of σ, is a subset S of X. If $\operatorname{dom}\sigma = X$, then σ is a map on X (or, for emphasis, a **total map**). The set of partial maps on X is denoted $(X \multimap X)$; this set contains all total maps on X and all partial determinations of them. The set $(X \multimap X)$ is ordered in the following way: given $\sigma, \tau \in (X \multimap X)$, define $\sigma \leqslant \tau$ if and only if $\operatorname{dom}\sigma \subseteq \operatorname{dom}\tau$ and $\sigma(x) = \tau(x)$ for all $x \in \operatorname{dom}\sigma$. Equivalently, $\sigma \leqslant \tau$ if and only if $\operatorname{graph}\sigma \subseteq \operatorname{graph}\tau$ in $\wp(X \times X)$. Note that a subset G of $X \times X$ is the graph of a partial map if and only if

$$(\forall s \in X)\,((s, x) \in G \;\&\; (s, x') \in G) \implies x = x'.$$

Now consider, for definiteness, the case $X = \mathbb{N}$. In a computation to determine a map $f\colon \mathbb{N} \to \mathbb{N}$, we may envisage information about f being output in stages. In the simplest situation, stage k of the computation might output $f(k)$. More generally, at each finite stage of the computation, the output would be some partial map σ, where $\sigma < f$ in the ordering defined above on $(\mathbb{N} \multimap \mathbb{N})$. We may think of f as being built up from tokens of information, each of which is an element of $(\mathbb{N} \multimap \mathbb{N})$ which partially specifies f. In the other direction, suppose we are given a collection $\mathcal{F}$ of elements of $(\mathbb{N} \multimap \mathbb{N})$. Is there a map f such that we have $\sigma \leqslant f$ for each $\sigma \in \mathcal{F}$? Clearly, the tokens must not supply conflicting messages about the putative f. For example, f cannot exist if $\mathcal{F}$ contains elements σ and τ such that, for some $n \in \mathbb{N}$, we have $(n, m) \in \operatorname{graph}\sigma$ and $(n, m') \in \operatorname{graph}\tau$, where $m \neq m'$. We say that a subset $\mathcal{F}$ of $(\mathbb{N} \multimap \mathbb{N})$ is **consistent** if, for any finite subset, $\mathcal{G}$, of $\mathcal{F}$, there exists $\rho \in (X \multimap X)$ (but not necessarily in $\mathcal{F}$) such that $\sigma \leqslant \rho$ for all $\sigma \in \mathcal{G}$. It is easy to see that, so long as $\mathcal{F}$ is a consistent subset of $(\mathbb{N} \multimap \mathbb{N})$, there exists a map $f\colon \mathbb{N} \to \mathbb{N}$ such that $\sigma \leqslant f$ for all $\sigma \in \mathcal{F}$. The concept of consistency is treated in a more general setting in Chapter 3.

The examples above illustrate ways in which ordered sets can model situations in which the relation $x \leqslant y$ has interpretations such as 'y is more defined than x' or 'y is a better approximation than x'. In each case, we have a notion of a **total object** (a completely defined, or idealized, element). These total objects are the 1-point intervals in the first example, the infinite binary strings in the second, and the total maps

in the third. The most interesting examples from a computational point
of view are those in which the total objects may be realized as limits (in
an order-theoretic sense to be investigated in Chapter 3) of objects which
are in some sense finite. A finite object should be one which encodes a
finite amount of information: for example, finite strings, or partial maps
with finite domains. These issues are taken up in Chapter 3.

1.7 Ordered sets in the humanities and social sciences. Below is a
pot-pourri of examples to indicate how ordered sets occur in the social
sciences and elsewhere. Each of these areas of application has led to the
investigation of ordered sets of special types.

An **interval order** on a set X is an order relation such that there
is a mapping φ of the points of X into subintervals of $\mathbb{R}$ such that,
for $x < y$ in X, the right-hand endpoint of $\varphi(x)$ is less than the left-
hand endpoint of $\varphi(y)$. Interval orders model, for example, the time
spans over which animal species are found or the occurrence of styles of
pottery in archaeological strata. A variant on the definition requires all
the image intervals to be of the same length, with problems of inexact
measurement in mind.

The problem of amalgamating the expressed preferences of a group
of individuals to arrive at a consensus is of concern to selection commit-
tees, market researchers, psephologists and many others. More explicitly,
given m objects and rankings of them by n individuals specified by n
chains, how should a chain be constructed which best reflects the individuals' collective preferences? A **social choice function** assigns to any
n-tuple of rankings a single ranking which defines a consensus, according
to specified criteria. A famous theorem, due to K. Arrow, asserts that
there is a set of criteria which are very natural but mutually incompat-
ible. This paradoxical result set off an avalanche of research on social
choice theory.

An increasingly important area of mathematics deals with the construction of algorithms for solving scheduling problems. Many such problems involve precedence constraints. For example, certain stages in the
assembly of a car must precede others and a conference organizer is likely
to have to schedule certain lectures before others. The computational
complexity of a scheduling problem depends critically on the order relation which describes the precedence constraints.

Order enters into the classification of objects on two rather different levels. The first is illustrated by our introductory example of the
arrangement of the planets into a hierarchical list according to their distance from the sun and by Figure 3.1 which classifies certain ordered sets
according to various criteria. On a deeper level, the rather new discipline

of **concept analysis** provides a powerful technique for classifying and for analyzing complex sets of data. From a set of objects (to take a simple example, the planets) and a set of attributes (for the planets, perhaps large/small, moon/no moon, near sun/far from sun), concept analysis builds an ordered set which reveals inherent hierarchical structure and thence natural groupings and dependencies among the objects and the attributes. Chapter 11 gives a brief introduction to concept analysis.

Diagrams

One of the most useful and attractive features of ordered sets is that, in the finite case at least, they can be 'drawn'. To describe how to represent ordered sets diagrammatically, we need the idea of covering.

1.8 The covering relation. Let P be an ordered set and let $x, y \in P$. We say x is **covered by** y (or y **covers** x), and write $x \prec y$ or $y \succ x$, if $x < y$ and $x \leqslant z < y$ implies $z = x$. The latter condition is demanding that there be no element z of P with $x < z < y$.

Observe that, if P is finite, $x < y$ if and only if there exists a finite sequence of covering relations $x = x_0 \prec x_1 \prec \ldots \prec x_n = y$. Thus, in the finite case, the order relation determines, and is determined by, the covering relation.

Here are some simple examples.

(i) In the chain $\mathbb{N}$, we have $m \prec n$ if and only if $n = m + 1$.

(ii) In $\mathbb{R}$, there are no pairs x, y such that $x \prec y$.

(iii) In $\wp(X)$, we have $A \prec B$ if and only if $B = A \cup \{b\}$, for some $b \in X \setminus A$.

1.9 Diagrams. Let P be a finite ordered set. We can represent P by a configuration of circles (representing the elements of P) and interconnecting lines (indicating the covering relation). The construction goes as follows.

(i) To each point $x \in P$, associate a point $P(x)$ of the euclidean plane $\mathbb{R}^2$, depicted by a small circle with centre at $P(x)$.

(ii) For each covering pair $x \prec y$ in P, take a line segment $\ell(x, y)$ joining the circle at $P(x)$ to the circle at $P(y)$.

(iii) Carry out (i) and (ii) in such a way that

    (a) if $x \prec y$, then $P(x)$ is 'lower' than $P(y)$ (that is, in standard cartesian coordinates, has a strictly smaller second coordinate);

    (b) the circle at $P(z)$ does not intersect the line segment $\ell(x, y)$ if $z \neq x$ and $z \neq y$.

It is easily proved by induction on the size, $|P|$, of P that (iii) can be achieved. A configuration satisfying (i)–(iii) is called a **diagram** (or **Hasse diagram**) of P. In the other direction, a diagram may be used to define a finite ordered set; an example is given below. Of course, the same ordered set may have many different diagrams. Diagram-drawing is as much an art as a science, and, as will become increasingly apparent as we proceed, good diagrams can be a real asset to understanding and to theorem-proving.

Figure 1.1(i) shows two alternative diagrams for the ordered set $P = \{a, b, c, d\}$ in which $a < c$, $a < d$, $b < c$ and $b < d$. (When we specify an ordered set by a set of inequalities in this way it is to be understood that no other pairs of distinct elements are comparable.) In Figure 1.1(ii) we have drawings which are not legitimate diagrams for P; in the first (iii)(a) in 1.9 is violated and in the second (iii)(b) is.

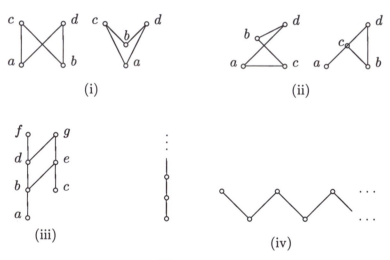

Figure 1.1

It is easy to tell from a diagram whether one element of an ordered set is less than another: $x < y$ if and only if there is a sequence of connected line segments moving upwards from x to y. Thus in the ordered set defined by the diagram in Figure 1.1(iii), $e \parallel f$ and $a < g$, for example.

We have only defined diagrams for finite ordered sets. It is not possible to represent the whole of an infinite ordered set by a diagram, but if its structure is sufficiently regular it can often be suggested diagrammatically, as indicated by the examples in Figure 1.1(iv).

1.10 Examples. Figure 1.2 contains diagrams for a variety of ordered sets. All possible ordered sets with three elements are presented in (i). In (ii) we have diagrams for **2**, **4** and $\overline{\mathbf{3}}$.

Figure 1.2(iii)(a) depicts $\wp(\{1, 2, 3\})$ (known as the **cube**). A less satisfying, but equally valid, diagram for the same ordered set is shown in Figure 1.2(iii)(b). In Figure 1.2(iv) are diagrams for $\operatorname{Sub} G$ for $G = \mathbf{V}_4$, the Klein 4-group, and $G = \mathbf{S}_3$, the symmetric group on 3 letters; in each case the subset $\mathcal{N}\text{-}\operatorname{Sub} G$ is shaded.

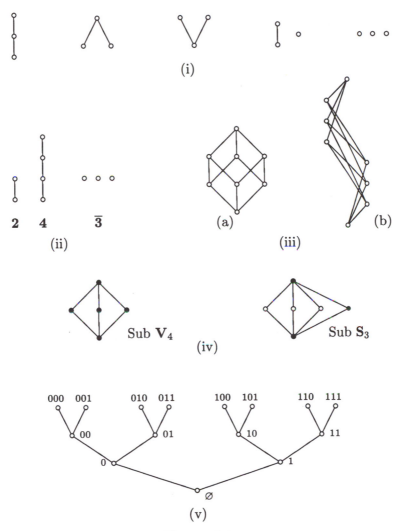

Figure 1.2

Figure 1.2(v) gives a diagram for the subset of Σ^* consisting of strings of length not more than 3.

Maps between ordered sets

This section introduces structure-preserving maps between ordered sets. In particular, it provides the machinery for deciding when two ordered sets are essentially the same.

1.11 Definitions. Let P and Q be ordered sets. A map $\varphi\colon P \to Q$ is said to be

(i) **order-preserving** (or, alternatively, **monotone**) if $x \leqslant y$ in P implies $\varphi(x) \leqslant \varphi(y)$ in Q;

(ii) an **order-embedding** if $x \leqslant y$ in P if and only if $\varphi(x) \leqslant \varphi(y)$ in Q;

(iii) an **order-isomorphism** if it is an order-embedding mapping P onto Q.

When $\varphi\colon P \to Q$ is an order-embedding we write $\varphi\colon P \hookrightarrow Q$. When there exists an order-isomorphism from P to Q, we say that P and Q are **order-isomorphic** and write $P \cong Q$.

1.12 Examples. Figure 1.3 shows some maps between ordered sets. The map φ_1 is not order-preserving. Each of φ_2–φ_5 is order-preserving, but not an order-embedding. The map φ_6 is an order-embedding, but not an order-isomorphism.

1.13 Remarks. The following are all easy to prove.

(1) Let $\varphi\colon P \to Q$ and $\psi\colon Q \to R$ be order-preserving maps. Then the composite map $\psi \circ \varphi$, given by $(\psi \circ \varphi)(x) = \psi(\varphi(x))$ for $x \in P$, is order-preserving. More generally the composite of a finite number of order-preserving maps is order-preserving, if it is defined.

(2) Let $\varphi\colon P \hookrightarrow Q$ and let $\varphi(P)$ (defined to be $\{\, \varphi(x) \mid x \in P \,\}$) be the image of φ. Then $\varphi(P) \cong P$. This justifies the use of the term embedding.

(3) An order-embedding is automatically a one-to-one map (because $\leqslant$ is reflexive on Q and antisymmetric on P). An order-isomorphism is bijective (that is, one-to-one and onto).

(4) Ordered sets P and Q are order-isomorphic if and only if there exist order-preserving maps $\varphi\colon P \to Q$ and $\psi\colon Q \to P$ such that

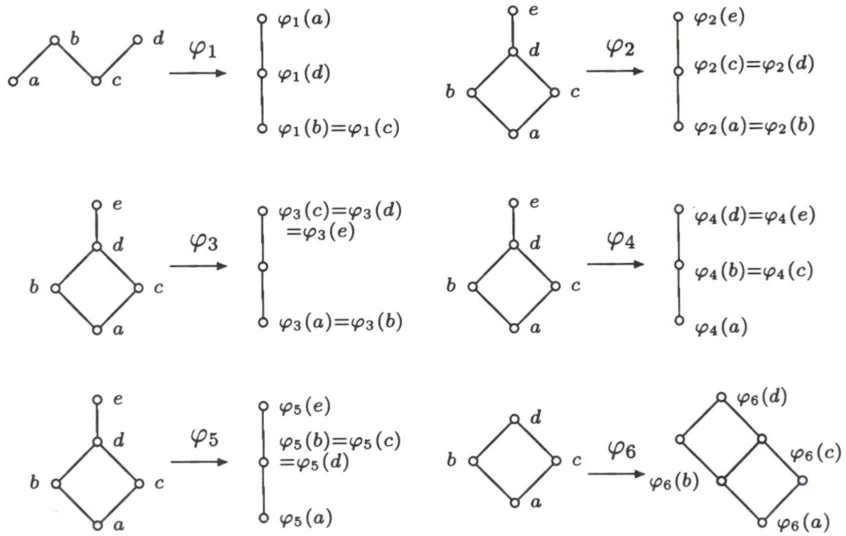

Figure 1.3

$\varphi \circ \psi = \mathrm{id}_Q$ and $\psi \circ \varphi = \mathrm{id}_P$ (where $\mathrm{id}_S \colon S \to S$ denotes the **identity map** on S given by $\mathrm{id}_S(x) = x$ for all $x \in S$.)

The diagrammatic approach to finite ordered sets is made fully legitimate by Proposition 1.15, which follows easily from Lemma 1.14.

1.14 Lemma. *Let P and Q be finite ordered sets and let $\varphi \colon P \to Q$ be a bijective map. Then the following are equivalent:*

(i) *φ is an order-isomorphism;*

(ii) *$x < y$ in P if and only if $\varphi(x) < \varphi(y)$ in Q;*

(iii) *$x \prec y$ in P if and only if $\varphi(x) \prec \varphi(y)$ in Q.*

Proof. The equivalence of (i) and (ii) is immediate from the definitions.

Now assume (ii) holds and take $x \prec y$ in P. Then $x < y$, so $\varphi(x) < \varphi(y)$ in Q. Suppose there exists $w \in Q$ such that $\varphi(x) < w < \varphi(y)$. Since φ is onto, there exists $u \in P$ such that $w = \varphi(u)$. By (ii), $x < u < y \,\lightning$. Hence $\varphi(x) \prec \varphi(y)$. The reverse implication is proved in much the same way. Hence (iii) holds.

Now assume (iii) and let $x < y$ in P. Then there exist elements $x = x_0 \prec x_1 \prec \ldots \prec x_n = y$. By (iii), $\varphi(x_0) = \varphi(x) \prec \varphi(x_1) \prec \ldots \prec \varphi(x_n) = \varphi(y)$. Hence $\varphi(x) < \varphi(y)$. The reverse implication is proved similarly, using the fact that φ is onto. Hence (ii) holds. ∎

1.15 Proposition. *Two finite ordered sets P and Q are order-isomorphic if and only if they can be drawn with identical diagrams.*

Proof. Assume there exists an order-isomorphism $\varphi \colon P \to Q$. To show that the same diagram represents both P and Q, note that the diagram is determined by the covering relation and invoke 1.14 (i) $\Rightarrow$ (iii). Conversely, assume P and Q can both be represented by the same diagram, D. Then there exist bijective maps f and g from P and Q onto the points of D. The composite map $\varphi = g^{-1} \circ f$ is bijective and satisfies condition (iii) in Lemma 1.14, so is an order-isomorphism.   ∎

1.16 Speaking categorically.   In modern pure mathematics it is rare for a class of structures of a given type to be introduced without an associated class of structure-preserving maps following hard on its heels. Ordered sets + order-preserving maps is one example. Others are groups + group homomorphisms, vector spaces over a field + linear maps, topological spaces + continuous maps, and we later meet lattices + lattice homomorphisms, CPOs + continuous maps, etc., etc. The recognition that an appropriate unit for study is a class of objects together with its structure-preserving maps (or **morphisms**) leads to category theory. Informally, a **category** is a class of objects + morphisms, with an operation of composition of morphisms satisfying a set of natural conditions suggested by examples such as those above.

Commuting diagrams of objects and morphisms expressing properties of categories and *their* structure-preserving maps (called **functors**) form the basis of category theory. The representation theory we present in Chapters 8 and 10 is a prime example of a topic which owes its development to the apparatus of category theory. We do not have sufficient need to call on the theory of categories to warrant setting up its formalism here, but it would be wrong not to acknowledge its subliminal influence.

The Duality Principle; down-sets and up-sets

1.17 The dual of an ordered set.   Given any ordered set P we can form a new ordered set P^{∂} (the **dual** of P) by defining $x \leqslant y$ to hold in P^{∂} if and only if $y \leqslant x$ holds in P. For P finite, we obtain a diagram for P^{∂} simply by 'turning upside down' a diagram for P. Figure 1.4 provides a simple illustration.

To each statement about P there corresponds a statement about P^{∂}. For example, we can assert that in P in Figure 1.4 there exists a unique element covered by exactly one other element, while in P^{∂} there exists a unique element covering exactly one other element. In general, given any statement Φ about ordered sets, we obtain the **dual statement** Φ^{∂} by replacing each occurrence of $\leqslant$ by $\geqslant$ and vice versa.

Thus ordered set concepts and results hunt in pairs. This fact can often be used to give two theorems for the price of one or to reduce work (as, for example, in the proof of Theorem 5.2). The formal basis for this observation is the Duality Principle below; its proof is a triviality.

P $\qquad\qquad\qquad\qquad P^\partial$

Figure 1.4

1.18 The Duality Principle. *Given a statement Φ about ordered sets which is true in all ordered sets, then the dual statement Φ^∂ is true in all ordered sets.*

Associated with any ordered set are two important families of sets.

1.19 Definitions and remarks. Let P be an ordered set and $Q \subseteq P$.

(i) Q is a **down-set** (alternative terms include **decreasing set** or **order ideal**) if, whenever $x \in Q$, $y \in P$ and $y \leqslant x$, we have $y \in Q$.

(ii) Dually, Q is an **up-set** (alternative terms are **increasing set** or **order filter**) if, whenever $x \in Q$, $y \in P$ and $y \geqslant x$, we have $y \in Q$.

It may help to think of a down-set as one which is 'closed under going down'. Down-sets and up-sets may be depicted in a stylized way in a 'directional Venn diagram'; see Figure 1.5. Such drawings do not have the formal status of diagrams, as defined in 1.9.

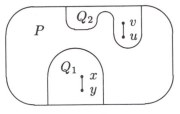 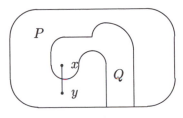

Q_1 a down-set, Q_2 an up-set $\qquad\qquad$ Q not a down-set

Figure 1.5

Besides being related by duality, down-sets and up-sets are related by complementation: Q is a down-set if and only if $P \setminus Q$ is an up-set. The proof is left as an exercise.

Given an arbitrary subset Q of P and $x \in P$, we define

$$\downarrow Q = \{y \in P \mid (\exists x \in Q)\, y \leqslant x\} \quad \text{and} \quad \uparrow Q = \{y \in P \mid (\exists x \in Q)\, y \geqslant x\},$$
$$\downarrow x = \{y \in P \mid y \leqslant x\} \quad \text{and} \quad \uparrow x = \{y \in P \mid y \geqslant x\}.$$

These are read 'down Q', etc. It is easily checked that $\downarrow Q$ is the smallest down-set containing Q and that Q is a down-set if and only if $Q = \downarrow Q$, and dually for $\uparrow Q$. Clearly $\downarrow\{x\} = \downarrow x$, and dually.

The family of all down-sets of P is denoted by $\mathcal{O}(P)$. It is itself an ordered set, under the inclusion order, and plays a crucial role in later chapters. The letter $\mathcal{O}$ is traditional here; it comes from the term *order ideal*.

When P is finite, every non-empty set in $\mathcal{O}(P)$ is expressible in the form $\bigcup_{i=1}^{k} \downarrow x_i$ (where $\{x_1, \ldots, x_k\}$ is an antichain). This provides a recipe for finding $\mathcal{O}(P)$ (though one which is practical only when P is small).

1.20 Examples.

(1) Consider the ordered set in Figure 1.1(iii). The sets $\{c\}$, $\{a, b, c, d, e\}$ and $\{a, b, d, f\}$ are all down-sets. The set $\{b, d, e\}$ is not a down-set; we have $\downarrow\{b, d, e\} = \{a, b, c, d, e\}$. The set $\{e, f, g\}$ is an up-set, but $\{a, b, d, f\}$ is not.

(2) Figure 1.6 shows $\mathcal{O}(P)$ in a simple case.

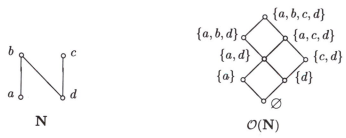

Figure 1.6

(3) If P is an antichain, then $\mathcal{O}(P) = \wp(P)$.

(4) If P is the chain $\mathbf{n}$, then $\mathcal{O}(P)$ consists of all the sets $\downarrow x$ for $x \in P$, together with the empty set. Hence $\mathcal{O}(P)$ is an $(n + 1)$-element chain. If P is the chain of rational numbers, $\mathbb{Q}$, then $\mathcal{O}(P)$

contains the empty set, $\mathbf{Q}$ itself and all sets $\downarrow x$ (for $x \in \mathbf{Q}$). There are other sets in $\mathcal{O}(P)$ too: for example, $\downarrow x \smallsetminus \{x\}$ (for $x \in \mathbf{Q}$) and $\{y \in \mathbf{Q} \mid y < a\}$ (for $a \in \mathbf{R} \smallsetminus \mathbf{Q}$).

We conclude this section with a useful lemma connecting the order relation and down-sets. The proof is an easy but instructive exercise.

1.21 Lemma. *Let P be an ordered set and $x, y \in P$. Then the following are equivalent:*

(i) $x \leqslant y$;

(ii) $\downarrow x \subseteq \downarrow y$;

(iii) $(\forall Q \in \mathcal{O}(P)) \, y \in Q \implies x \in Q$.

Maximal and minimal elements; top and bottom

We next introduce some important special elements.

1.22 Maximal and minimal elements. Let P be an ordered set and let $Q \subseteq P$. Then

(i) $a \in Q$ is a **maximal** element of Q if $a \leqslant x \in Q$ implies $a = x$;

(ii) $a \in Q$ is the **greatest** (or **maximum**) element of Q if $a \geqslant x$ for every $x \in Q$, and in that case we write $a = \max Q$.

Figure 1.7

A **minimal** element of Q, the **least** (or **minimum**) element of Q and $\min Q$ are defined dually, that is by reversing the order. Observe that Q has a greatest element only if it has precisely one maximal element, and that the greatest element of Q, if it exists, is unique (by the antisymmetry of $\leqslant$). Any non-empty subset of a finite ordered set P always has at least one maximal element. Figure 1.7 illustrates the definitions: P_1 has maximal elements a_1, a_2, a_3, but no greatest element; a is the greatest element of P_2.

A subset of the chain $\mathbf{N}$ has a maximal element if and only if it is finite. In the subset of $\wp(\mathbf{N})$ consisting of all subsets of $\mathbf{N}$ except $\mathbf{N}$ itself, there is no greatest element, but $\mathbf{N} \smallsetminus \{n\}$ is maximal for each

$n \in \mathbb{N}$. The subset of $\wp(\mathbb{N})$ consisting of all finite subsets of $\mathbb{N}$ has no maximal elements. These examples confirm that an ordered set may have a unique maximal element, many maximal elements, or none. An important set-theorists' tool, known as **Zorn's Lemma**, guarantees the existence of maximal elements, under suitable conditions. Zorn's Lemma is discussed in Chapter 4.

Referring to the examples in 1.6, we see that the maximal elements in Σ^{**} are the infinite strings and those in $(X \multimap X)$ are the total maps. This suggests that when an order relation models information we might expect a correlation between maximal elements and totally defined elements.

1.23 Top and bottom. Let P be an ordered set. The greatest element of P, if it exists, is called the **top element** of P and written $\top$ (pronounced 'top'). Similarly, the least element of P, if such exists, is called the **bottom element** and denoted $\bot$ ('bottom').

In $\wp(X)$, we have $\top = X$ and $\bot = \varnothing$. A finite chain always has top and bottom elements, but an infinite chain need not have. For example, the chain $\mathbb{N}$ has bottom element 1, but no top, while $\mathbb{Z}$ possesses neither top nor bottom. Top and bottom do not exist in any antichain with more than one element.

In the context of information orderings, $\bot$ and $\top$ have the following interpretations: $\bot$ represents 'no information', while $\top$ corresponds to an over-determined, or contradictory, element. None of the ordered sets in 1.6 has a top element. Each has a bottom element: this is $[-\infty, \infty]$ for interval approximations to real numbers, the empty string for Σ^{**} and the partial map with empty domain for $(X \multimap X)$. In each case $\bot$ is the least informative element. In modelling computations, a bottom element is also useful for representing and handling non-termination. Accordingly, computer scientists commonly choose as models ordered sets which have bottom elements, but prefer their ordered sets topless.

1.24 Lifting. In Chapters 3 and 4 almost all the results refer to ordered sets with $\bot$. Lack of a bottom element can be easily remedied by adding one. Given an ordered set P (with or without $\bot$), we form $P_\bot$ (called P 'lifted') as follows. Take an element $\mathbf{0} \notin P$ and define $\leqslant$ on $P_\bot := P \cup \{\mathbf{0}\}$ by

$$x \leqslant y \text{ if and only if } x = \mathbf{0} \text{ or } x \leqslant y \text{ in } P.$$

Any set S gives rise to an ordered set with $\bot$, as follows. Order S by making it an antichain, $\overline{S}$, and then form $\overline{S}_\bot$. Ordered sets obtained in this way are called **flat**. In applications it is likely that $S \subseteq \mathbb{R}$. In this context we shall for simplicity write $S_\perp$ instead of the more correct $\overline{S}_\perp$. Since we shall not have occasion to apply lifting to subsets of $\mathbb{R}$ ordered as chains this should cause no confusion. Figure 1.8 shows $\mathbb{N}_\perp$.

Figure 1.8

Building new ordered sets

This section collects together a number of ways of constructing new ordered sets from existing ones. The other way round, it will often be helpful to analyze ordered sets by regarding them as built up from simpler components.

For completeness we should add to the catalogue below those constructions already introduced. Given an ordered set P,

(i) any subset Q of P becomes an ordered set with the induced order;

(ii) the dual ordered set P^∂ is obtained by reversing the order on P;

(iii) $P_\perp$ is constructed by adjoining a (new) bottom element to P.

Where we refer to diagrams in the sequel, it is to be assumed that the ordered sets involved are finite.

1.25 Sums of ordered sets. There are several different ways to join two ordered sets together. In each of these constructions we require that the sets being joined are disjoint (and we shall assume this for the remainder of this subsection). This is no real restriction since we can always find isomorphic copies of the original ordered sets which are disjoint; see Exercise 1.21 for a formal approach to this process.

Suppose that P and Q are (disjoint) ordered sets. The **disjoint union** $P \mathbin{\dot\cup} Q$ of P and Q is the ordered set formed by defining $x \leqslant y$ in $P \cup Q$ if and only if either $x, y \in P$ and $x \leqslant y$ in P or $x, y \in Q$ and $x \leqslant y$ in Q. A diagram for $P \mathbin{\dot\cup} Q$ is formed by placing side by side diagrams for P and Q.

Again let P and Q be (disjoint) ordered sets. The **linear sum** $P \oplus Q$ is defined by taking the following order relation on $P \cup Q$: $x \leqslant y$ if and only if

$$x, y \in P \text{ and } x \leqslant y \text{ in } P,$$
$$\text{or} \quad x, y \in Q \text{ and } x \leqslant y \text{ in } Q,$$
$$\text{or} \quad x \in P, y \in Q.$$

A diagram for $P \oplus Q$ is obtained by placing a diagram for P directly below a diagram for Q and then adding a line segment from *each* maximal element of P to *each* minimal element of Q. The lifting construction is a special case of a linear sum: $P_\perp$ is just $\mathbf{1} \oplus P$. Similarly, $P \oplus \mathbf{1}$ represents P with a (new) top element added.

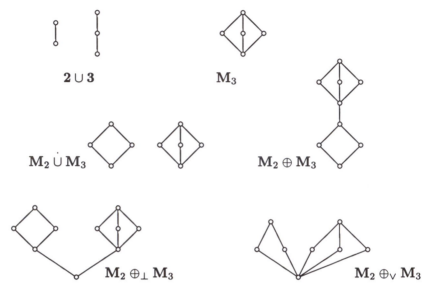

2 $\dot{\cup}$ 3   **M$_3$**

M$_2$ $\dot{\cup}$ M$_3$ **M$_2$ $\oplus$ M$_3$**

M$_2$ $\oplus_\perp$ M$_3$ **M$_2$ $\oplus_\vee$ M$_3$**

Figure 1.9

Each of the operations $\dot{\cup}$ and $\oplus$ is associative: for (pairwise disjoint) ordered sets P, Q and R,

$$P \dot{\cup} (Q \dot{\cup} R) = (P \dot{\cup} Q) \dot{\cup} R \text{ and } P \oplus (Q \oplus R) = (P \oplus Q) \oplus R.$$

This allows us to write iterated disjoint and linear sums unambiguously without brackets. We denote by $\mathbf{M}_n$ the sum $\mathbf{1} \oplus \overline{\mathbf{n}} \oplus \mathbf{1}$.

Let P and Q be disjoint ordered sets each with a bottom element. Then $P \dot{\cup} Q$ fails to have a bottom element. There are two ways to modify this construction and stay within the class of ordered sets with $\perp$. The first is to form $(P \dot{\cup} Q)_\perp$; this is the **separated sum** of P and Q, which we write as $P \oplus_\perp Q$. Alternatively, we may form the **coalesced sum**, $P \oplus_\vee Q$, by taking $P \dot{\cup} Q$ and identifying the two bottom elements.

Figure 1.9 shows examples of sums.

1.26 Products. Let $P_1, \ldots, P_n$ be ordered sets. The cartesian product $P_1 \times \cdots \times P_n$ can be made into an ordered set by imposing the coordinatewise order defined by

$$(x_1, \ldots, x_n) \leqslant (y_1, \ldots, y_n) \iff (\forall i) \, x_i \leqslant y_i \text{ in } P_i.$$

Given an ordered set P, the notation P^n is used as shorthand for the n–fold product $P \times \cdots \times P$.

As an aside we remark that there is another way to order the product of ordered sets P and Q. Define the **lexicographic order** by $(x_1, x_2) \leqslant (y_1, y_2)$ if $x_1 < y_1$ or ($x_1 = y_1$ and $x_2 \leqslant y_2$). By iteration a lexicographic order can be defined on any finite product of ordered sets. Unless otherwise stated we shall always equip a product with the coordinatewise order.

Informally, a product $P \times Q$ is drawn by replacing each point of a diagram of P by a copy of a diagram for Q, and connecting 'corresponding' points; this assumes that the points are placed in such a way that the rules for diagram-drawing in 1.9 are obeyed. Figure 1.10(i) shows diagrams for some simple products. In Figure 1.10(ii) we depict the four-dimensional hypercube $\mathbf{2}^4$ in various ways. The third representation is obtained by thinking of $\mathbf{2}^4$ as order-isomorphic to $\mathbf{2} \times \mathbf{2}^3$.

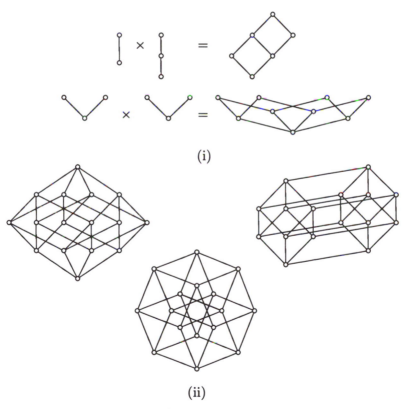

(i)

(ii)

Figure 1.10

The diagram for 2^3 is the same as that for $\wp(\{1,2,3\})$ in Figure 1.2(iii). By 1.15, these two ordered sets are isomorphic. We can prove directly a more general result.

1.27 Proposition. Let $X = \{1, 2, \ldots, n\}$ and define $\psi \colon \wp(X) \to 2^n$ by $\psi(A) = (\varepsilon_1, \ldots, \varepsilon_n)$ where

$$\varepsilon_i = \begin{cases} 1 & \text{if } i \in A, \\ 0 & \text{if } i \notin A. \end{cases}$$

Then ψ is an order-isomorphism.

Proof. Given $A, B \in \wp(X)$, let $\psi(A) = (\varepsilon_1, \ldots, \varepsilon_n)$ and $\varphi(B) = (\delta_1, \ldots, \delta_n)$. Then

$$\begin{aligned} A \subseteq B &\iff (\forall i)\, i \in A \text{ implies } i \in B \\ &\iff (\forall i)\, \varepsilon_i = 1 \text{ implies } \delta_i = 1 \\ &\iff (\forall i)\, \varepsilon_i \leqslant \delta_i \\ &\iff \psi(A) \leqslant \psi(B) \text{ in } 2^n. \end{aligned}$$

Hence ψ is an order-embedding. Take $x = (\varepsilon_1, \ldots, \varepsilon_n) \in 2^n$. Then $x = \psi(A)$, where $A = \{i \mid \varepsilon_i = 1\}$, so ψ is onto. ∎

1.28 Ordered sets of maps. In elementary analysis an ordering of functions is used and understood often without a formal definition ever being given. Consider, for example, the statement '$\sin x \leqslant x^2$ on $\mathbf{R}$'. The order relation implicit here is the **pointwise order**: for functions $f, g \colon \mathbf{R} \to \mathbf{R}$, $f \leqslant g$ means $f(x) \leqslant g(x)$ for all $x \in \mathbf{R}$.

Pointwise ordering need not be confined to real-valued functions on $\mathbf{R}$. Suppose X is any set and Y an ordered set. We may order the set Y^X of all maps from X to Y as follows. We put $f \leqslant g$ if and only if $f(x) \leqslant g(x)$ in Y, for all $x \in X$. When X is an n-element set, then Y^X is really just Y^n, as defined in 1.26.

Any subset Q of Y^X inherits the pointwise order. When X is itself an ordered set, we may take Q to be the set of all order-preserving maps from X to Y; the resulting ordered set is denoted $Y^{\langle X \rangle}$. We sometimes write $(X \to Y)$ in place of Y^X and $\langle X \to Y \rangle$ in place of $Y^{\langle X \rangle}$. This alternative notation is needed because the notation Y^X and $Y^{\langle X \rangle}$ becomes unwieldy when X or Y is of the form $P_\perp$ or when higher-order functions are involved. A higher-order function means a function which maps functions to functions; a typical example is the map ψ in 1.29.

1.29 Partial maps again. We can use the lifting process to give new insight into the partial maps introduced in 1.6. Let S be any set, in particular a subset of $\mathbf{R}$, and define $S_\perp$ as in 1.24. With each $\pi \in (S \multimap S)$ we associate a map $\psi(\pi) \colon S \to S_\perp$, given by $\psi(\pi) = \pi_\perp$ where

$$\pi_\perp(x) = \begin{cases} \pi(x) & \text{if } x \in \operatorname{dom} \pi, \\ \mathbf{0} & \text{otherwise.} \end{cases}$$

Recall that $\mathbf{0}$ is $\perp$ in $S_\perp$. Thus ψ is a map from $(S \multimap S)$ to $(S \to S_\perp)$; we have used the extra element $\mathbf{0}$ to convert a partial map on S into an 'ordinary' map defined on the whole of S.

We claim that ψ sets up an order-isomorphism between $(S \multimap S)$ and $(S \to S_\perp)$. For $\pi, \sigma \in (S \multimap S)$,

$$\psi(\pi) \leqslant \psi(\sigma) \text{ in } S \to S_\perp$$
$$\Longleftrightarrow (\forall x \in S)(\pi_\perp(x) \leqslant \sigma_\perp(x))$$
$$\Longleftrightarrow (\forall x \in S)(\pi_\perp(x) = \mathbf{0} \text{ or } \pi_\perp(x) = \sigma_\perp(x))$$
$$\Longleftrightarrow (\forall x \in S)(x \in \operatorname{dom} \pi \Rightarrow x \in \operatorname{dom} \sigma \text{ and } \pi(x) = \sigma(x))$$
$$\Longleftrightarrow \pi \leqslant \sigma \text{ in } (S \multimap S);$$

here the first equivalence uses the definition of $\leqslant$ in $(S \to S_\perp)$ and the second uses the definition of $\leqslant$ in $S_\perp$. Hence ψ is an order-embedding. To show that ψ is onto, take $f \colon S \to S_\perp$. Let $T = \{\, x \in S \mid f(x) \neq \mathbf{0} \,\}$ and define $\pi \colon T \to S$ by $\pi(x) = f(x)$ for $x \in T$. Then $\psi(\pi) = f$. Hence ψ is an order-isomorphism. ∎

Exercises

Exercises from the text. Verify the unproved assertions in 1.13 and 1.19. Prove Lemma 1.21.

1.1 Let P be a set on which a binary relation $<$ is defined such that, for all $x, y, z \in P$,

(a) $x < x$ is false,

(b) $x < y$ and $y < z$ imply $x < z$.

Prove that if $\leqslant$ is defined by

$$x \leqslant y \Longleftrightarrow (x < y \text{ or } x = y),$$

then $\leqslant$ is an order on P, and moreover every order on P arises from a relation $<$ satisfying (a) and (b). [A binary relation satisfying (a) and (b) is called a **strict order**.]

1.2 There is a list of 16 diagrams of four-element ordered sets such that every four-element ordered set can be represented by one of

the diagrams in the list. (That is, up to order-isomorphism, there
are just 16 four-element ordered sets.) Find such a list.

1.3   Recall from 1.4 that $\preccurlyeq$ is defined on (any subset of) N_0 by $m \preccurlyeq n$ if
and only if m divides n. Draw a diagram for each of the following
subsets of $\langle \mathsf{N}_0; \preccurlyeq \rangle$:

    (a) $\{1, 2, 3, 5, 30\}$,   (b) $\{1, 2, 3, 4, 12\}$,
    (c) $\{1, 2, 5, 10\}$,   (d) $\{1, 2, 4, 8, 16\}$,
    (e) $\{2, 3, 12, 18\}$,   (f) $\{1, 2, 3, 4, 6, 12\}$,
    (g) $\{1, 2, 3, 12, 18, 0\}$,   (h) $\{1, 2, 3, 5, 6, 10, 15, 30\}$.

1.4   Let $P = \{a, b, c, d, e, f, u, v\}$. Draw the diagram of the ordered set
$\langle P; \leqslant \rangle$ where

$$v < a, \ v < b, \ v < c, \ v < d, \ v < e, \ v < f, \ v < u,$$

$$a < c, \ a < d, \ a < e, \ a < f, \ a < u,$$

$$b < c, \ b < d, \ b < e, \ b < f, \ b < u,$$

$$c < d, \ c < e, \ c < f, \ c < u,$$

$$d < e, \ d < f, \ d < u,$$

$$e < u, \ f < u.$$

1.5   In which of the following cases is the map $\varphi \colon P \to Q$ order-
preserving?

    (i) $P = Q = \langle \mathbb{Z}; \leqslant \rangle$, and $\varphi(x) = x + 1$.

   (ii) $P = \langle \wp(S); \subseteq \rangle$ where $|S| > 1$, $Q = \mathbf{2}$, and $\varphi(U) = 1$ if $U \neq$
$\varnothing$, $\varphi(U) = 0$ if $U = \varnothing$.

  (iii) $P = \langle \wp(S); \subseteq \rangle$ where $|S| > 1$, $Q = \mathbf{2}$, and $\varphi(U) = 1$ if $U =$
S, $\varphi(U) = 0$ if $U \neq S$.

  (iv) $P = Q = \langle \mathsf{N}_0; \preccurlyeq \rangle$, and $\varphi(x) = nx$ (with $n \in \mathsf{N}_0$ fixed).

   (v) $P = \langle \mathcal{O}(S); \subseteq \rangle$, $Q = \mathbf{2}$, and $\varphi(U) = 1$ if $x \in U$, $\varphi(U) =$
0 otherwise (with $x \in S$ fixed).

  (vi) $P = Q = \langle \wp(\mathsf{N}); \subseteq \rangle$, and φ defined by

$$\varphi(U) = \begin{cases} \{1\} & \text{if } 1 \in U, \\ \{2\} & \text{if } 2 \in U \text{ and } 1 \notin U, \\ \varnothing & \text{otherwise.} \end{cases}$$

1.6   Let $\mathsf{N}_0^* := \{ n \in \mathsf{N} \mid 2 \text{ does not divide } n \} \cup \{0\}$.

    (i) Define φ from $\langle \mathsf{N}_0; \preccurlyeq \rangle$ to $\langle \mathsf{N}_0^*; \preccurlyeq \rangle$ by $\varphi(0) := 0$ and otherwise
$\varphi(n) := n/2^k$, where 2^k is the highest power of 2 which

divides n. Show that φ is order-preserving and maps onto $\mathbb{N}_0^*$ but is not an order-isomorphism.

(ii) Show that $\langle \mathbb{N}_0; \preccurlyeq \rangle$ is order-isomorphic to $\langle \mathbb{N}_0^*; \preccurlyeq \rangle$.

1.7 Prove that the ordered set Σ^{**} of all binary strings is a **tree** (that is, an ordered set P with $\perp$ such that $\downarrow x$ is a chain for each $x \in P$). For each $u \in \Sigma^{**}$ describe the set of elements covering u.

1.8 Let P be the set of all finite binary strings (including the empty string). Define $\leqslant$ on P by $u \leqslant v$ if and only if v is a prefix of u or there exist (possibly empty) strings x, y, z such that $v = x0y$ and $u = x1z$. Show that $\leqslant$ is an order on P and that $\langle P; \leqslant \rangle$ is a chain with a $\top$ but no $\perp$. Draw a diagram of the induced order on the seven strings of length less than three and another for the fifteen strings of length less than four.

Let Q be the set of all finite or infinite binary strings with $\leqslant$ defined as for P; then again $\langle Q; \leqslant \rangle$ is a chain. Does Q have a $\perp$? Show that if u is an infinite string then (i) there is no string v such that $u \prec v$, and (ii) there is a string w such that $w \prec u$ if and only if u contains only a finite number of zeros. [Hint. First consider the particular cases $u_1 = 101010\overline{1}$ and $u_2 = \overline{10}$, where $\overline{x} = xxx \dots$ for any finite string x.]

1.9 Let A and B be down-sets of P. Prove that $A \prec B$ in $\langle \mathcal{O}(P); \subseteq \rangle$ if and only if $B = A \cup \{b\}$ for some minimal element b of $X \setminus A$.

1.10 Draw and label a diagram for $\mathcal{O}(P)$ for each of the ordered sets P of Figure 1.11.

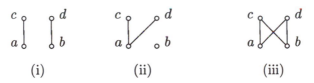

(i)   (ii)   (iii)

Figure 1.11

1.11 Let $F_n = \{x_1, \dots, x_n\}$ with $x_1 < x_2 > x_3 < \dots x_n$ or $x_1 > x_2 < x_3 > \dots x_n$ (and no other comparabilities); then F_n is called a **fence**. Recall that the Fibonnaci sequence $1, 1, 2, 3, 5, 8 \dots$ is defined by $f_1 = f_2 = 1$ and $f_k = f_{k-2} + f_{k-1}$ for all $k \geqslant 3$. Show that $|\mathcal{O}(F_n)| = f_{n+2}$ for all $n \geqslant 1$.

1.12 Let P and Q be ordered sets.

   (i) Show that $\varphi\colon P \to Q$ is order-preserving if and only if
       $\varphi^{-1}(A)$ is a down-set in P whenever A is a down-set in Q.

  (ii) Assume $\varphi\colon P \to Q$ is order-preserving. Then, by (i), the
       map $\varphi^{-1}\colon \mathcal{O}(Q) \to \mathcal{O}(P)$ is well defined.

      (a) Show that φ is an order-embedding if and only if φ^{-1}
          maps $\mathcal{O}(Q)$ onto $\mathcal{O}(P)$.

      (b) Show that φ maps onto Q if and only if $\varphi^{-1}\colon \mathcal{O}(Q) \to$
          $\mathcal{O}(P)$ is one-to-one.

1.13 Draw the diagrams of the products shown in Figure 1.12.

(i)   (ii)

Figure 1.12

1.14 Let P and Q be ordered sets. Prove that $(a_1, b_1) \prec (a_2, b_2)$ in
      $P \times Q$ if and only if

$$(a_1 = a_2 \ \& \ b_1 \prec b_2) \quad \text{or} \quad (a_1 \prec a_2 \ \& \ b_1 = b_2).$$

1.15 Prove that, for all ordered sets P, Q and R,

$$\langle P \to \langle Q \to R \rangle \rangle \cong \langle P \times Q \to R \rangle.$$

1.16 Let X be any set and let Y be an ordered set. Show that $f \prec g$
      in Y^X if and only if there exists $x_0 \in X$ such that

      (a) $f(x) = g(x)$ for all $x \in X \setminus \{x_0\}$,

      (b) $f(x_0) \prec g(x_0)$.

      Assume that X is a finite ordered set. Show that $f \prec g$ in $Y^{\langle X \rangle}$
      if and only if there exists $x_0 \in X$ such that (a) and (b) hold.
      [Hint. The 'if' direction follows from part (i). To prove the 'only
      if' direction, argue by contradiction. Assume that $f \prec g$ and
      suppose that f and g differ at more than one element of X. Choose
      $y \in X$ minimal with respect to $f(y) < g(y)$ and let $z \in X \setminus \{y\}$
      satisfy $f(z) < g(z)$. Define $h\colon X \to Y$ by

$$h(x) = \begin{cases} f(x) & \text{if } x = y, \\ g(x) & \text{if } x \neq y. \end{cases}$$

Use the minimality of y to prove that h is order-preserving and then show that $f < h < g$. Finally, show that if $f \prec g$ and $f(x_0) < g(x_0)$, then $f(x_0) \prec g(x_0)$.]

1.17 Draw diagrams for $P^{\langle 2 \rangle}$ and $P^{\langle 3 \rangle}$ where P is **2**, **3** or $V := \mathbf{1} \oplus (\mathbf{1} \,\dot{\cup}\, \mathbf{1})$. Label the elements—an element of $P^{\langle 2 \rangle}$ may be labelled xy where $x \leqslant y$ in P and similarly an element of $P^{\langle 3 \rangle}$ may be labelled xyz where $x \leqslant y \leqslant z$ in P.

1.18 Let P and Q be chains. Prove that $P \times Q$ is a chain in the lexico-graphic order. Prove that $P \times Q$ is a chain in the coordinatewise order if and only if at most one of P and Q has more than one element.

1.19 Let P, Q and R be ordered sets. Is it true that $P \odot (Q \odot R) \cong (P \odot Q) \odot R$ when $\odot$ denotes the operation of forming (i) the separated sum, (ii) the coalesced sum?

1.20 Let P and Q be ordered sets. Describe, in terms of down-sets of P and Q, the down-sets of the following: $P_\perp$, $P \,\dot{\cup}\, Q$, $P \oplus Q$, $P \times Q$ with the lexicographic order.

1.21 Let P and Q be ordered sets with $P \cap Q \neq \varnothing$. Give formal definitions of the ordered sets $P \,\dot{\cup}\, Q$ and $P \oplus Q$. [Hint. Define appropriate orders on $(\{0\} \times P) \cup (\{1\} \times Q)$.]

1.22 Let $\rho \subseteq P \times P$ be a binary relation on a set P. Then the **transitive closure**, ρ^t, of ρ is defined by $a \, \rho^t \, b$ if and only if

$$(\exists n \in \mathbb{N})(\exists z_0, z_1, \ldots, z_n \in P)\, a = z_0 \, \rho \, z_1 \, \rho \, z_2 \ldots z_{n-1} \, \rho \, z_n = b.$$

(i) Let $\leqslant$ be an order on P and assume that $a \parallel b$ for some $a, b \in P$. Show that ρ^t is an order on P where

$$\rho^t := \{\,(x, y) \mid x \leqslant y\,\} \cup \{(a, b)\}.$$

[Hint. While this can be done directly, it is easier to work with the corresponding strict order, $<$, and show that the transitive closure of

$$\{\,(x, y) \mid x < y\,\} \cup \{(a, b)\}$$

is also a strict order. See Exercise 1.1.]

(ii) Use (i) to show that if P is finite then every order $\leqslant$ on P has a **linear extension**, that is, there is an order $\leqslant_1$ such that

$\langle P; \leqslant_1 \rangle$ is a chain and for all $a, b \in P$ we have $a \leqslant b$ implies $a \leqslant_1 b$. [The infinite case will be proved in Exercise 4.22.]

(iii) Use (i) to show that if $\langle P; \leqslant \rangle$ is a finite ordered set, then there is a finite number of chains $\langle P; \leqslant_1 \rangle, \ldots, \langle P; \leqslant_n \rangle$ such that for all $a, b \in P$ we have

$$a \leqslant b \iff (a \leqslant_1 b \ \& \ldots \& \ a \leqslant_n b).$$

(iv) Show that $\leqslant_1$ is a linear extension of the order $\leqslant$ on P if and only if $\langle P; \leqslant_1 \rangle$ is a chain and the identity map $\mathrm{id} \colon P \to P$ is an order-preserving map from $\langle P; \leqslant \rangle$ to $\langle P; \leqslant_1 \rangle$.

(v) Draw and label a diagram for every possible linear extension of the ordered set $\langle \mathbf{N}; \leqslant \rangle$ given in Figure 1.6.

1.23 Let P be a finite ordered set. The **width** of P is defined to be the size of the largest antichain in P and is denoted by $w(P)$.

(i) Find $w(P)$ for each of the ordered sets P in Figure 1.11 and show that in each case that P can be written as a union of $w(P)$ many chains.

(ii) Show that if a finite ordered set P can be written as the union of n chains, then $n \geqslant w(P)$.

(iii) **Dilworth's Theorem** states that the width $w(P)$ of a finite ordered set P equals the least $n \in \mathbf{N}$ such that P can be written as a union of n chains. The more intrepid may try to find their own proof of this important result. Alternatively, a much easier, but still valuable, exercise is to rewrite the snappy 14–line proof from [34], explaining every step in detail.

1.24 Show that it is impossible to find an order $\leqslant$ on $\mathbf{C}$ such that for all $w, z \in \mathbf{C}$

(a) $z = 0$ or $z > 0$ or $z < 0$,

(b) $w, z > 0$ imply $(-w) < 0$, $w + z > 0$ and $wz > 0$.

1.25 Let X be a topological space satisfying

(T_0): given $x \neq y$ in X there exists either an open set U such that $x \in U$, $y \notin U$ or an open set V such that $x \notin V$, $y \in V$.

Show that $\leqslant$, defined by $x \leqslant y$ if and only if $x \in \overline{\{y\}}$, is an order on X. Describe the down-sets and the up-sets for this order.

2

Lattices and Complete Lattices

Many important properties of an ordered set P are expressed in terms of the existence of certain upper bounds or lower bounds of subsets of P. Two of the most important classes of ordered sets defined in this way are lattices and complete lattices.

Lattices as ordered sets

It is a fundamental property of the real numbers, $\mathbb{R}$, that if I is a closed and bounded interval in $\mathbb{R}$, then every subset of I has both a least upper bound (or supremum) and a greatest lower bound (or infimum) in I. These concepts pertain to any ordered set.

2.1 Definitions. Let P be an ordered set and let $S \subseteq P$. An element $x \in P$ is an **upper bound** of S if $s \leqslant x$ for all $s \in S$. A **lower bound** is defined dually. The set of all upper bounds of S is denoted by S^u (read as 'S **upper**') and the set of all lower bounds by S^ℓ (read as 'S **lower**'):

$$S^u := \{\, x \in P \mid (\forall s \in S)\, s \leqslant x \,\} \text{ and } S^\ell := \{\, x \in P \mid (\forall s \in S)\, s \geqslant x \,\}.$$

Since $\leqslant$ is transitive, S^u is always an up-set and S^ℓ a down-set. If S^u has a least element, x, then x is called the **least upper bound** of S. Equivalently, x is the least upper bound of S if

(i) x is an upper bound of S, and

(ii) $x \leqslant y$ for all upper bounds y of S.

Dually, if S^ℓ has a largest element, x, then x is called the **greatest lower bound** of S. Since least elements and greatest elements are unique (see 1.22), least upper bounds and greatest lower bounds are unique when they exist. The least upper bound of S is also called the **supremum** of S and is denoted by $\sup S$; the greatest lower bound of S is also called the **infimum** of S and is denoted by $\inf S$.

2.2 Remark. The two extreme cases, where S is empty or S is P itself, warrant a brief investigation. Recall from 1.23 that, when they exist, the top and bottom elements of P are denoted by $\top$ and $\bot$ respectively. It is easily seen that if P has a top element, then $P^u = \{\top\}$ in which case $\sup P = \top$. When P has no top element, we have $P^u = \varnothing$ and hence $\sup P$ does not exist. By duality, $\inf P = \bot$ whenever P has a bottom element. Now let S be the empty subset of P. Then every element

$x \in P$ satisfies (vacuously) $s \leqslant x$ for all $s \in S$. Thus $\varnothing^u = P$ and hence $\sup \varnothing$ exists if and only if P has a bottom element, and in that case $\sup \varnothing = \bot$. Dually, $\inf \varnothing = \top$ whenever P has a top element.

2.3 Notation. Looking ahead to Chapter 5, we shall adopt the following neater notation: we write $x \vee y$ (read as 'x **join** y') in place of $\sup\{x, y\}$ when it exists and $x \wedge y$ (read as 'x **meet** y') in place of $\inf\{x, y\}$ when it exists. Similarly we write $\bigvee S$ (the '**join of** S') and $\bigwedge S$ (the '**meet of** S') instead of $\sup S$ and $\inf S$ when these exist. It is sometimes necessary to indicate that the join or meet is being found in a particular ordered set P, in which case we write $\bigvee_P S$ or $\bigwedge_P S$.

2.4 Remarks.

(1) Let P be any ordered set. If $x, y \in P$ and $x \leqslant y$, then $\{x, y\}^u = {\uparrow}y$ and $\{x, y\}^\ell = {\downarrow}x$. Since the least element of ${\uparrow}y$ is y and the greatest element of ${\downarrow}x$ is x, we have $x \vee y = y$ and $x \wedge y = x$ whenever $x \leqslant y$. In particular, since $\leqslant$ is reflexive, we have $x \vee x = x$ and $x \wedge x = x$.

(2) In an ordered set P, the least upper bound, $x \vee y$, of $\{x, y\}$ may fail to exist for two different reasons:

(a) because x and y have no common upper bound, or

(b) because they have no *least* upper bound.

In Figure 2.1(a) we have $\{a, b\}^u = \varnothing$ and hence $a \vee b$ doesn't exist. In (b) we find that $\{a, b\}^u = \{c, d\}$ and thus $a \vee b$ doesn't exist as $\{a, b\}^u$ has no least element.

(a)   (b)

Figure 2.1

(3) Consider the ordered set drawn in Figure 2.2. It is tempting, at first sight, to think that $b \vee c = i$. On more careful inspection we find that $\{b, c\}^u = \{\top, h, i\}$. Since $\{b, c\}^u$ has distinct minimal elements, namely h and i, it cannot have a least element; hence $b \vee c$ does not exist. On the other hand, $\{a, b\}^u = \{\top, h, i, f\}$ has a least element, namely f, and thus $a \vee b = f$.

We shall be particularly interested in ordered sets in which $x \vee y$ and $x \wedge y$ exist for all $x, y \in P$.

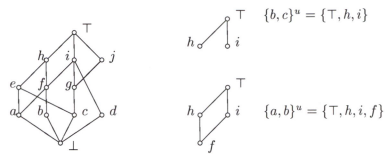

Figure 2.2

2.5 Definitions. Let P be a non-empty ordered set.

(i) If $x \vee y$ and $x \wedge y$ exist for all $x, y \in P$, then P is called a **lattice**.

(ii) If $\bigvee S$ and $\bigwedge S$ exist for all $S \subseteq P$, then P is called a **complete lattice**.

If P is a lattice, then $\vee$ and $\wedge$ are binary operations on P and we have an algebraic structure $\langle P; \vee, \wedge \rangle$. This theme will be developed in Chapter 5.

2.6 Remarks.

(1) In 2.5(ii), $S = \varnothing$ is allowed. Hence, by 2.2, any complete lattice is **bounded**, that is, has top and bottom elements.

(2) Let P be a non-empty ordered set. By Remark 2.4(1), if $x \leqslant y$ then $x \vee y = y$ and $x \wedge y = x$. Hence to show that P is a lattice it suffices to prove that $x \vee y$ and $x \wedge y$ exist in P for all non-comparable pairs $x, y \in P$.

2.7 Examples.

(1) By 2.6(2), every chain is a lattice in which $x \vee y = \max\{x, y\}$ and $x \wedge y = \min\{x, y\}$. Thus each of $\mathbb{R}, \mathbb{Q}, \mathbb{Z}$ and $\mathbb{N}$ is a lattice under its usual order. None of them is complete; every one lacks a top element. However, if $-\infty < x < y < \infty$, then the closed interval $[x, y]$ in $\mathbb{R}$ is a complete lattice (by the completeness axiom for $\mathbb{R}$). Failure of completeness in $\mathbb{Q}$ is more fundamental than in $\mathbb{R}$. In $\mathbb{Q}$, it is not merely the lack of top and bottom elements which causes problems; for example, the set $\{\, s \in \mathbb{Q} \mid s^2 < 2 \,\}$ has upper bounds but no least upper bound in $\mathbb{Q}$.

(2) For any set X, the ordered set $\wp(X)$ is a complete lattice in which

$$\bigvee \{\, A_i \mid i \in I \,\} = \bigcup_{i \in I} A_i,$$

$$\bigwedge \{\, A_i \mid i \in I \,\} = \bigcap_{i \in I} A_i.$$

Let $\mathfrak{L} \subseteq \wp(X)$. Then $\mathfrak{L}$ is known as a **lattice of sets** if it is closed under finite unions and intersections and a **complete lattice of sets** if it is closed under arbitrary unions and intersections. If $\mathfrak{L}$ is a lattice of sets, then $\langle \mathfrak{L}; \subseteq \rangle$ is a lattice in which $A \vee B = A \cup B$ and $A \wedge B = A \cap B$. Similarly, if $\mathfrak{L}$ is a complete lattice of sets, then $\langle \mathfrak{L}; \subseteq \rangle$ is a complete lattice with join given by set union and meet given by set intersection.

Let P be an ordered set and consider the ordered set $\mathcal{O}(P)$ of all down-sets of P introduced in 1.19. If $\{A_i\}_{i \in I} \subseteq \mathcal{O}(P)$, then $\bigcup_{i \in I} A_i$ and $\bigcap_{i \in I} A_i$ both belong to $\mathcal{O}(P)$. Hence $\mathcal{O}(P)$ is a complete lattice of sets.

(3) The ordered set $\mathbf{M}_n$ (for $n \geqslant 1$) introduced in Chapter 1 (see Figure 2.3) is easily seen to be a lattice. Let $x, y \in \mathbf{M}_n$ with $x \parallel y$. Then x and y are in the central antichain of $\mathbf{M}_n$ and hence $x \vee y = \top$ and $x \wedge y = \bot$.

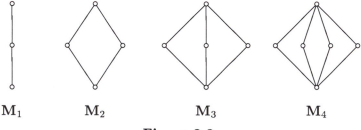

$\mathbf{M}_1 \qquad\qquad \mathbf{M}_2 \qquad\qquad\qquad \mathbf{M}_3 \qquad\qquad\qquad \mathbf{M}_4$

Figure 2.3

(4) Consider the ordered set $\langle \mathbb{N}_0; \preccurlyeq \rangle$ of non-negative integers ordered by division (see 1.4). Recall that k is the greatest common divisor (or highest common factor) of m and n if
   (a) k divides both m and n (that is, $k \preccurlyeq m$ and $k \preccurlyeq n$),
   (b) if ℓ divides both m and n, then ℓ divides k (that is, $\ell \preccurlyeq k$ for all $k \in \{m, n\}^{\ell}$).

Thus the greatest common divisor of m and n is precisely the meet of m and n in $\langle \mathbb{N}_0; \preccurlyeq \rangle$. Dually, the join of m and n in $\langle \mathbb{N}_0; \preccurlyeq \rangle$ is given by their least common multiple. You should convince yourself that these statements remain valid when m or n equals 0. Thus $\langle \mathbb{N}_0; \preccurlyeq \rangle$ is a lattice in which

$$m \vee n = \operatorname{lcm}\{m, n\} \quad \text{and} \quad m \wedge n = \gcd\{m, n\}.$$

Exercise 2.21 indicates two proofs that this lattice is complete.

2.8 Lattices of subgroups. Assume that G is a group and $\langle \operatorname{Sub} G; \subseteq \rangle$ its ordered set of subgroups. It is certainly true that $H \cap K \in \operatorname{Sub} G$, whence $H \wedge K$ exists and equals $H \cap K$. Tiresomely, $H \cup K$ is only exceptionally a subgroup. Nevertheless, $H \vee K$ does exist in $\operatorname{Sub} G$, as (rather tautologically) the subgroup $\langle H \cup K \rangle$ generated by $H \cup K$. Unfortunately there is no convenient general formula for $H \vee K$. This lopsided behaviour is analyzed more closely in the next section, where we consider arbitrary joins and meets in $\operatorname{Sub} G$.

Normal subgroups are more amenable. Meet is again given by $\cap$ and join in $\mathcal{N}\text{-Sub}\, G$ has a particularly compact description. It is a straightforward exercise in group theory to show that if H, K are normal subgroups of G, then

$$HK := \{\, hk \mid h \in H, \; k \in K \,\}$$

is also a normal subgroup of G. It follows easily that the join in $\mathcal{N}\text{-Sub}\, G$ is given by $H \vee K = HK$.

The lattices $\operatorname{Sub} G$ and $\mathcal{N}\text{-Sub}\, G$ are given in Figure 2.4 for the group, $\mathbf{D}_4$, of symmetries of a square and for the group $\mathbb{Z}_2 \times \mathbb{Z}_4$. The elements of $\mathcal{N}\text{-Sub}\, G$ are shaded. (You should convince yourself, from the diagrams, that these ordered sets are indeed lattices.)

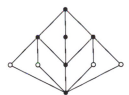 

$\operatorname{Sub} \mathbf{D}_4$ with $\mathcal{N}\text{-Sub}\, \mathbf{D}_4$ shaded $\qquad$ $\operatorname{Sub} \mathbb{Z}_2 \times \mathbb{Z}_4 = \mathcal{N}\text{-Sub}\, \mathbb{Z}_2 \times \mathbb{Z}_4$

Figure 2.4

We may conjecture that properties of a group G are reflected in properties of $\operatorname{Sub} G$ and $\mathcal{N}\text{-Sub}\, G$, and vice versa. To take a very simple example, a group G is finite if and only if $\operatorname{Sub} G$ is finite (an easy exercise). More interestingly, many group-theoretic properties of G are equivalent to order- or lattice-theoretic properties of $\operatorname{Sub} G$ or

$\mathcal{N}$-Sub G. Some results in this spirit can be found in the exercises for Chapters 5, 6 and 8. These results do not appear in the text because their proofs lean much more heavily on group theory than on lattice theory.

The following lemmas contain useful information for computing with joins and meets. The first is an immediate consequence of the definitions of least upper bound and greatest lower bound.

2.9 Lemma. *Let P be an ordered set, let $S, T \subseteq P$ and assume that $\bigvee S$, $\bigvee T$, $\bigwedge S$ and $\bigwedge T$ exist in P.*

(i) *For all $s \in S, s \leqslant \bigvee S$ and $s \geqslant \bigwedge S$.*

(ii) *Let $x \in P$; then $x \leqslant \bigwedge S$ if and only if $x \leqslant s$ for all $s \in S$.*

(iii) *Let $x \in P$; then $x \geqslant \bigvee S$ if and only if $x \geqslant s$ for all $s \in S$.*

(iv) *$\bigvee S \leqslant \bigwedge T$ if and only if $s \leqslant t$ for all $s \in S$ and all $t \in T$.*

(v) *If $S \subseteq T$, then $\bigvee S \leqslant \bigvee T$ and $\bigwedge S \geqslant \bigwedge T$.*

A straightforward application of this lemma yields the next one, which shows that join and meet behave well with respect to set union. We leave the proof as an exercise.

2.10 Lemma. *Let P be a lattice, let $S, T \subseteq P$ and assume that $\bigvee S$, $\bigvee T$, $\bigwedge S$ and $\bigwedge T$ exist in P. Then*

$$\bigvee (S \cup T) = \left(\bigvee S\right) \vee \left(\bigvee T\right) \quad \text{and} \quad \bigwedge (S \cup T) = \left(\bigwedge S\right) \wedge \left(\bigwedge T\right).$$

An easy induction now yields the following lemma.

2.11 Lemma. *Let P be a lattice. Then $\bigvee F$ and $\bigwedge F$ exist for every finite non-empty subset F of P.*

2.12 Corollary. *Every finite lattice is complete.*

The last of this group of elementary lemmas describes how order-preserving maps interact with joins and meets.

2.13 Lemma. *Let P and Q be ordered sets, $\varphi \colon P \to Q$ be an order-preserving map and assume that $S \subseteq P$ is such that $\bigvee S$ and $\bigwedge S$ exist in P and $\bigvee \varphi(S)$ and $\bigwedge \varphi(S)$ exist in Q. Then*

$$\varphi\left(\bigvee S\right) \geqslant \bigvee \varphi(S) \quad \text{and} \quad \varphi\left(\bigwedge S\right) \leqslant \bigwedge \varphi(S).$$

2.14 Combinations of lattices and complete lattices. In Chapter 1 we discussed a variety of ways to combine ordered sets: sums, products, etc. We now enquire which of these constructions produce (complete) lattices when applied to (complete) lattices.

The sum constructions in 1.25 rarely yield lattices, but it is true that the linear sum of lattices P and Q is again a lattice.

We next consider the product of lattices P and Q. For elements (x_1, y_1) and (x_2, y_2) of $P \times Q$, it is easily seen that $(x_1, y_1) \vee (x_2, y_2)$ exists and equals $(x_1 \vee x_2, y_1 \vee y_2)$, and similarly for meet. Hence $P \times Q$ is a lattice. Further, if P and Q are complete, then $P \times Q$ is complete, with joins and meets being formed coordinatewise.

Now assume that P is any set and Q a (complete) lattice. Then, under the usual pointwise order, the set Q^P of maps from P to Q is a (complete) lattice. Join and meet are obtained pointwise. This means that the join, φ, of $\{\varphi_i\}_{i \in I}$ is given by

$$(\forall x \in P)\, \varphi(x) = \bigvee_{i \in I} \varphi_i(x),$$

and similarly for meet. When P carries an order relation and all the maps φ_i are order-preserving, then $\bigvee \{\, \varphi_i \mid i \in I \,\}$ and $\bigwedge \{\, \varphi_i \mid i \in I \,\}$ are also order-preserving. Thus $Q^{\langle P \rangle}$ is a (complete) lattice.

The rest of this chapter deals with complete lattices. Chapters 3 and 4 continue the study of completeness conditions, and pursue applications to algebra, computer science and set theory. Completeness does not play a central role in Chapters 5–10. Some readers may now wish to proceed to Chapter 5, which develops the elementary algebraic theory of lattices, and to return to the intervening material later.

Complete lattices

To show that an ordered set is a complete lattice requires only half as much work as the definition would have us believe.

2.15 Lemma. *Let P be an ordered set such that $\bigwedge S$ exists in P for every non-empty subset S of P. Then $\bigvee S$ exists in P for every subset S of P which has an upper bound in P; indeed, $\bigvee S = \bigwedge S^u$.*

Proof. Let $S \subseteq P$ and assume that S has an upper bound in P; thus $S^u \neq \varnothing$. Hence, by assumption, $a := \bigwedge S^u$ exists in P. We claim that $\bigvee S = a$. The details are left as an exercise. ∎

2.16 Theorem.   *Let P be a non-empty ordered set. Then the following are equivalent:*

(i)   *P is a complete lattice;*

(ii) *$\bigwedge S$ exists in P for every subset S of P;*

(iii) *P has a top element, $\top$, and $\bigwedge S$ exists in P for every non-empty subset S of P.*

Proof. It is trivial that (i) implies (ii), and (ii) implies (iii) since the meet of the empty subset of P exists only if P has a top element (by 2.2). It follows easily from the previous lemma that (iii) implies (i); the details are left to the reader.   ∎

This theorem has a simple corollary which, nevertheless, yields many examples of complete lattices.

2.17 Corollary.   *Let X be a set and let $\mathfrak{L}$ be a family of subsets of X, ordered as usual by inclusion, and such that*

(a) *$\bigcap_{i \in I} A_i \in \mathfrak{L}$ for every non-empty family $\{A_i\}_{i \in I} \subseteq \mathfrak{L}$ and*

(b) *$X \in \mathfrak{L}$.*

Then $\mathfrak{L}$ is a complete lattice in which

$$\bigwedge \{\, A_i \mid i \in I \,\} = \bigcap_{i \in I} A_i,$$

$$\bigvee \{\, A_i \mid i \in I \,\} = \bigcap \{\, B \in \mathfrak{L} \mid \bigcup_{i \in I} A_i \subseteq B \,\}.$$

Proof. By Theorem 2.16, to show that $\langle \mathfrak{L}; \subseteq \rangle$ is a complete lattice it suffices to show that $\mathfrak{L}$ has a top element and that the meet of every non-empty subset of $\mathfrak{L}$ exists in $\mathfrak{L}$. By (b), $\mathfrak{L}$ has a top element, namely X. Let $\{A_i\}_{i \in I}$ be a non-empty subset of $\mathfrak{L}$; then (a) gives $\bigcap_{i \in I} A_i \in \mathfrak{L}$. Since $\bigcap_{i \in I} A_i \subseteq A_j$ for all $j \in I$, it follows that $\bigcap_{i \in I} A_i$ is a lower bound of $\{A_i\}_{i \in I}$. If $B \in \mathfrak{L}$ is a lower bound of $\{A_i\}_{i \in I}$, then $B \subseteq A_i$ for all $i \in I$ and hence $B \subseteq \bigcap_{i \in I} A_i$. Thus $\bigcap_{i \in I} A_i$ is the greatest lower bound of $\{A_i\}_{i \in I}$ in $\mathfrak{L}$, that is,

$$\bigwedge \{\, A_i \mid i \in I \,\} = \bigcap_{i \in I} A_i.$$

Thus $\langle \mathfrak{L}; \subseteq \rangle$ is a complete lattice. Since X is an upper bound of $\{A_i\}_{i \in I}$ in $\mathfrak{L}$, Lemma 2.15 gives

$$\begin{aligned}
\bigvee \{\, A_i \mid i \in I \,\} &= \bigwedge \{A_i \mid i \in I\}^u \\
&= \bigcap \{\, B \in \mathfrak{L} \mid (\forall i \in I)\, A_i \subseteq B \,\} \\
&= \bigcap \{\, B \in \mathfrak{L} \mid \bigcup_{i \in I} A_i \subseteq B \,\}. \qquad \blacksquare
\end{aligned}$$

2.18 Definitions. If $\mathfrak{L}$ is a non-empty family of subsets of X which satisfies Condition (a) of Corollary 2.17, then $\mathfrak{L}$ is called an **intersection structure** (or $\bigcap$**-structure**) on X. If $\mathfrak{L}$ also satisfies (b), we refer to it as a **topped intersection structure** on X. An alternative term is **closure system**; see 2.22.

Intersection structures which occur in computer science are usually topless while those in algebra are almost invariably topped. In complete lattices, $\mathfrak{L}$, of this type, the meet is just set intersection, but in general the join is not set union. This is illustrated in the examples which follow.

2.19 Examples.

(1) Consider $(X \longrightarrow\!\!\!\circ\, X)$, where X is any set. From the observations in 1.6 we saw that the map $\pi \mapsto \operatorname{graph} \pi$ is an order-embedding of $(X \longrightarrow\!\!\!\circ\, X)$ into $\wp(X \times X)$. Let $\mathfrak{L}$ be the family of subsets of $X \times X$ which are graphs of partial maps. To prove that $\mathfrak{L}$ is closed under intersections, use the characterization given in 1.6: if $S \subseteq X \times X$, then $S \in \mathfrak{L}$ if and only if $(s, x) \in S$ and $(s, x') \in S$ imply $x = x'$. Thus $\mathfrak{L}$ is an $\bigcap$-structure. It is not topped unless $|X| \leqslant 1$.

(2) Each of the following is a topped $\bigcap$-structure and so forms a complete lattice under inclusion.

  (a) the subgroups, $\operatorname{Sub} G$, of a group G;

  (b) the normal subgroups, $\mathcal{N}$-$\operatorname{Sub} G$, of a group G;

  (c) the equivalence relations on a set X;

  (d) the subspaces of a vector space;

  (e) the convex subsets of a real vector space;

  (f) the subrings of a ring;

  (g) the ideals of a ring.

  These families all belong to a class of $\bigcap$-structures we shall consider further in Chapter 3.

(3) The closed subsets of a topological space are closed under finite unions and finite intersections and hence form a lattice of sets in which $A \vee B = A \cup B$ and $A \wedge B = A \cap B$. In fact, the closed sets form a topped $\bigcap$-structure and consequently the lattice of closed sets is complete. The formulae for arbitrary (rather than finite) joins and meets given in 2.17 show that, in general, meet is given by intersection while the join of a family of closed sets is not their union but is obtained by forming the closure of their union.

(4) Since the open subsets of a topological space are closed under arbitrary union and include the empty set, the dual of 2.16 shows that

they form a complete lattice under inclusion. The 'dual' of 2.17 shows that join and meet are given by

$$\bigvee\{A_i \mid i \in I\} = \bigcup_{i \in I} A_i,$$

$$\bigwedge\{A_i \mid i \in I\} = \mathrm{Int}(\bigcap_{i \in I} A_i),$$

where $\mathrm{Int}(A)$ is the interior of A.

2.20 Closure operators. There is an intimate connection between the topped intersection structure, $\Gamma(X)$, consisting of the closed subsets of a topological space X and the closure operator, $^- : \wp(X) \to \wp(X)$, which maps a subset A of X to its closure $\overline{A}$. Namely,

$$\Gamma(X) = \{A \subseteq X \mid \overline{A} = A\}$$

and

$$\overline{A} = \bigcap\{B \in \Gamma(X) \mid A \subseteq B\}.$$

In fact, this connection has nothing to do with topology and exists for any topped $\bigcap$–structure.

Let X be a set. A map $C : \wp(X) \to \wp(X)$ is a **closure operator** on X if, for all $A, B \subseteq X$,

(a) $A \subseteq C(A)$,
(b) if $A \subseteq B$, then $C(A) \subseteq C(B)$,
(c) $C(C(A)) = C(A)$.

A subset A of X is called **closed** (with respect to C) if $C(A) = A$.

Every topped $\bigcap$–structure gives rise to a closure operator and conversely. The proof of the following theorem is left as an exercise.

2.21 Theorem. *Let C be a closure operator on a set X. Then the family*

$$\mathcal{L}_C := \{A \subseteq X \mid C(A) = A\}$$

of closed subsets of X is a topped $\bigcap$–structure and so forms a complete lattice, when ordered by inclusion, in which

$$\bigwedge\{A_i \mid i \in I\} = \bigcap_{i \in I} A_i,$$

$$\bigvee\{A_i \mid i \in I\} = C(\bigcup_{i \in I} A_i).$$

Conversely, given a topped $\bigcap$–structure $\mathfrak{L}$ on X the formula

$$C_{\mathfrak{L}}(A) := \bigcap \{\, B \in \mathfrak{L} \mid A \subseteq B \,\}.$$

defines a closure operator $C_{\mathfrak{L}}$ on X.

2.22 Remarks. The relationship between topped $\bigcap$–structures and closure operators is a bijective one: the closure operator induced by the topped $\bigcap$–structure $\mathfrak{L}_C$ is C itself, and, similarly, the topped $\bigcap$–structure induced by the closure operator $C_{\mathfrak{L}}$ is $\mathfrak{L}$; in symbols,

$$C_{(\mathfrak{L}_C)} = C \quad \text{and} \quad \mathfrak{L}_{(C_{\mathfrak{L}})} = \mathfrak{L}.$$

Thus, whether we work with a topped $\bigcap$–structure or the corresponding closure operator is a matter of convenience.

It is worth noting that every complete lattice arises (up to order-isomorphism) as a topped $\bigcap$–structure on some set or, equivalently, as the complete lattice of closed sets with respect to some closure operator. (See Exercise 2.9.)

2.23 Examples.

(1) Let G be a group. Then the closure operator corresponding to the $\bigcap$–structure $\mathrm{Sub}\, G$ maps a subset A of G to the subgroup $\langle A \rangle$ generated by A.

(2) Let V be a vector space over a field F and let $\mathrm{Sub}\, V$ be the complete lattice of linear subspaces of V. The corresponding closure operator on V maps a subset A of V to its linear span.

(3) Let P be an ordered set. It is easily seen that the map $\downarrow\colon \wp(P) \to \wp(P)$ introduced in 1.19 is a closure operator. The corresponding topped $\bigcap$–structure is the family $\mathcal{O}(P)$ of all down-sets of P.

Chain conditions and completeness

By Corollary 2.12, every finite lattice is complete. There are various finiteness conditions, of which 'P **is finite**' is the strongest, which will guarantee that a lattice is complete.

2.24 Definitions. Let P be an ordered set.

(i) If $C = \{c_0, c_1, \ldots, c_n\}$ is a finite chain in P with $|C| = n + 1$, then we say that the **length** of C is n.

(ii) P is said to have **length** n, written $\ell(P) = n$, if the length of the longest chain in P is n.

(iii) P is of **finite length** if it has length n for some $n \in \mathbb{N}_0$.

(iv) P has **no infinite chains** if every chain in P is finite.

(v) P satisfies the **ascending chain condition**, (ACC), if given any sequence $x_1 \leqslant x_2 \leqslant \ldots \leqslant x_n \leqslant \ldots$ of elements of P, there exists $k \in \mathbb{N}$ such that $x_k = x_{k+1} = \ldots$. The dual of the ascending chain condition is the **descending chain condition**, (DCC).

2.25 Examples.

(1) The lattices $\mathbf{M}_n$ of Figure 2.3 are of length 2. A lattice of finite length has no infinite chains and so satisfies both (ACC) and (DCC).

(2) The lattice $\langle \mathbb{N}_0; \preccurlyeq \rangle$ satisfies (DCC) but not (ACC).

(3) Consider the lattices in Figure 2.5. The chain $\mathbb{N}$ satisfies (DCC) but not (ACC), and, dually, $\mathbb{N}^\partial$ satisfies (ACC) but not (DCC). The lattice $\mathbf{1} \oplus (\bigcup_{n \in \mathbb{N}} \mathbf{n}) \oplus \mathbf{1}$ is the simplest example of a lattice which has no infinite chains but is not of finite length.

(4) The finiteness conditions in 2.24 first arose in 'classical' algebra. For exampe, it can be shown that a vector space V is finite dimensional if and only if $\operatorname{Sub} V$ is of finite length, in which case $\dim V = \ell(\operatorname{Sub} V)$.

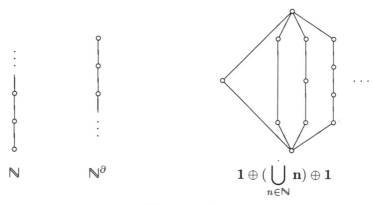

$$\mathbb{N} \qquad\qquad \mathbb{N}^\partial \qquad\qquad \mathbf{1} \oplus \left(\bigcup_{n \in \mathbb{N}} \mathbf{n} \right) \oplus \mathbf{1}$$

Figure 2.5

A formal proof of the following lemma requires some form of the axiom of set theory known as the **Axiom of Choice**. We give an informal derivation here and comment in Chapter 4 on how this can be converted into a formal proof (see 4.20).

2.26 Lemma. *An ordered set P satisfies (ACC) if and only if every non-empty subset A of P has a maximal element.*

Informal proof: We shall prove the contrapositive in both directions, that is, we prove that P has an infinite ascending chain if and only if there is a non-empty subset A of P which has no maximal element.

Assume that $x_1 < x_2 < \cdots < x_n < \ldots$ is an infinite ascending chain in P; then clearly $A := \{ x_n \mid n \in \mathbb{N} \}$ has no maximal element. Conversely, assume that A is a non-empty subset of P which has no maximal element. Let $x_1 \in A$. Since x_1 is not maximal in A, there exists $x_2 \in A$ with $x_1 < x_2$. Similarly, there exists $x_3 \in A$ with $x_2 < x_3$. Continuing in this way (and this is where the Axiom of Choice comes in) we obtain an infinite ascending chain in P. ∎

2.27 Theorem. *An ordered set P has no infinite chains if and only if it satisfies both* (ACC) *and* (DCC).

Proof. Clearly if P has no infinite chains, then it satisfies both (ACC) and (DCC). Suppose that P satisfies both (ACC) and (DCC) and contains an infinite chain C. Note that if A is a non-empty subset of C, then A has a maximal element, m, by 2.26. If $a \in A$, then since C is a chain we have $a \leqslant m$ or $m \leqslant a$. But $m \leqslant a$ implies $m = a$ by the maximality of m. Hence $a \leqslant m$ for all $a \in A$, and so every non-empty subset of C has a greatest element. Let x_1 be the greatest element of C, let x_2 be the greatest element of $C \setminus \{x_1\}$, and in general let x_{n+1} be the greatest element of $C \setminus \{x_1, x_2, \ldots, x_n\}$. Then $x_1 \succ x_2 \succ \ldots \succ x_n \succ \ldots$ is an infinite, descending, covering chain in P, ϟ. ∎

Lattices with no infinite chains are complete as the following more general result shows.

2.28 Theorem. *Let P be a lattice.*

(i) *If P satisfies* (ACC), *then for every non-empty subset A of P there exists a finite subset F of A such that $\bigvee A = \bigvee F$ (which exists in P by 2.11).*

(ii) *If P has a bottom element and satisfies* (ACC), *then P is complete.*

(iii) *If P has no infinite chains, then P is complete.*

Proof. Assume that P satisfies (ACC) and let A be a non-empty subset of P. Then, by 2.11,

$$B := \{ \bigvee F \mid F \text{ is a finite non-empty subset of } A \}$$

is a well-defined subset of P. Since B is non-empty, 2.26 guarantees that B has a maximal element $m = \bigvee F$ for some finite subset F of A. Let $a \in A$. Then $\bigvee(F \cup \{a\}) \in B$ and $m = \bigvee F \leqslant \bigvee(F \cup \{a\})$ by 2.9(v). Thus $m = \bigvee F = \bigvee(F \cup \{a\})$ since m is maximal in B. As

$m = \bigvee(F \cup \{a\})$ we have $a \leqslant m$, whence m is an upper bound of A. Let $x \in P$ be an upper bound of A. Then x is an upper bound of F since $F \subseteq A$ and hence $m = \bigvee F \leqslant x$. Thus m is the least upper bound of A; that is, $\bigvee A = m = \bigvee F$. Hence (i) holds.

Combining (i) with the dual of 2.16 yields (ii), and since a lattice with no infinite chains has a bottom element and satisfies (ACC), (iii) follows from (ii). ∎

Completions

In general, there are many ways in which an ordered set can be embedded into a complete lattice. In this section we shall concentrate on one such embedding—one which generalizes Dedekind's construction of $\mathbb{R}$ as the completion by cuts of $\mathbb{Q}$.

2.29 Definition and remarks. Let P be an ordered set. If $\varphi \colon P \hookrightarrow L$ and L is a complete lattice, then we say that L is a **completion** of P (via the order-embedding φ).

It is easily seen (Exercise 2.9) that the map $\varphi : x \mapsto {\downarrow} x$ is an order-embedding of P into $\mathcal{O}(P)$. We saw in 2.23(3) that $\mathcal{O}(P)$ is a complete lattice; hence $\mathcal{O}(P)$ is a completion of P. This completion is unnecessarily large. For example, if P is a complete lattice then P is a completion of itself (via the identity map) while $\mathcal{O}(P)$ is much larger.

Let P be an ordered set and let $A \subseteq P$. Recall from Definition 2.1 that A 'upper' and A 'lower' are defined by

$$A^u := \{\, x \in P \mid (\forall a \in A)\, a \leqslant x \,\} \quad \text{and} \quad A^\ell := \{\, x \in P \mid (\forall a \in A)\, a \geqslant x \,\}.$$

2.30 Lemma. *Let A and B be subsets of an ordered set P. Then*

(i) $A \subseteq A^{u\ell}$ *and* $A \subseteq A^{\ell u}$;

(ii) *if* $A \subseteq B$, *then* $A^u \supseteq B^u$ *and* $A^\ell \supseteq B^\ell$;

(iii) $A^u = A^{u\ell u}$ *and* $A^\ell = A^{\ell u \ell}$.

Proof. We have $a \leqslant x$ for all $a \in A$ and all $x \in A^u$, which says precisely that $A \subseteq (A^u)^\ell = A^{u\ell}$. Dually, $A \subseteq A^{\ell u}$. Thus (i) holds.

Clearly (ii) holds: if $A \subseteq B$, then any element of B^u is an upper bound of B and so is an upper bound of A and hence belongs to A^u.

By (i) we have $A \subseteq A^{u\ell}$, whence (ii) yields $A^u \supseteq (A^{u\ell})^u = A^{u\ell u}$. But (i), applied to A^u, also gives $A^u \subseteq (A^u)^{\ell u} = A^{u\ell u}$. Hence $A^u = A^{u\ell u}$ and, by duality $A^\ell = A^{\ell u \ell}$, which proves (iii). ∎

2.31 The Dedekind–MacNeille Completion. It follows very easily from Lemma 2.30 that $C(A) := A^{u\ell}$ defines a closure operator on P. By 2.21,

$$\mathbf{DM}(P) := \{\, A \subseteq P \mid A^{u\ell} = A \,\}$$

is a topped $\bigcap$-structure on P. Hence, by 2.16, $\langle \mathbf{DM}(P); \subseteq \rangle$ is a complete lattice, known as the **Dedekind–MacNeille completion** of P. (It is also referred to as the **completion by cuts** or the **normal** completion of P.)

2.32 Lemma. *Let P be an ordered set.*

(i) *For all $x \in P$, $(\downarrow x)^{u\ell} = \downarrow x$ and hence $\downarrow x \in \mathbf{DM}(P)$.*

(ii) *If $A \subseteq P$ and $\bigwedge A$ exists in P, then*

$$\bigcap \{\, \downarrow a \mid a \in A \,\} = \downarrow(\bigwedge A).$$

(iii) *If $A \subseteq P$ and $\bigvee A$ exists in P, then $A^{u\ell} = \downarrow(\bigvee A)$.*

Proof. (i) Let $y \in (\downarrow x)^u$; then $z \leqslant y$ for all $z \in \downarrow x$ so, in particular, $x \leqslant y$ as $x \in \downarrow x$ and hence $y \in \uparrow x$. Thus $(\downarrow x)^u \subseteq \uparrow x$. If $y \in \uparrow x$, then $y \geqslant x$ and so, by transitivity, $y \geqslant z$ for all $z \in \downarrow x$, that is, $y \in (\downarrow x)^u$. Thus $\uparrow x \subseteq (\downarrow x)^u$. Therefore $(\downarrow x)^u = \uparrow x$ and, by duality, $(\uparrow x)^\ell = \downarrow x$. Thus $(\downarrow x)^{u\ell} = (\uparrow x)^\ell = \downarrow x$.

(ii) Let $A \subseteq P$ and assume that $\bigwedge A$ exists in P. Note that

$$\bigcap \{\, \downarrow a \mid a \in A \,\} = \{\, x \in P \mid (\forall a \in A)\, x \leqslant a \,\} = A^\ell.$$

Since $\bigwedge A$ is a lower bound of A we have $\bigwedge A \in A^\ell$ and hence $\downarrow(\bigwedge A) \subseteq A^\ell$ as A^ℓ is a down set. Since $\bigwedge A$ is the *greatest* lower bound of A we have $x \leqslant \bigwedge A$ for all $x \in A^\ell$ and thus $A^\ell \subseteq \downarrow(\bigwedge A)$. Hence $A^\ell = \downarrow(\bigwedge A)$.

(iii) Let $A \subseteq P$ and assume that $\bigvee A$ exists in P. Of course $\bigvee A \in A^u$. Thus $x \in A^{u\ell}$ implies that $x \leqslant \bigvee A$ and hence $x \in \downarrow(\bigvee A)$. Consequently $A^{u\ell} \subseteq \downarrow(\bigvee A)$. Since $\bigvee A$ is the *least* upper bound of A we have $\bigvee A \leqslant y$ for all $y \in A^u$ and hence $\bigvee A \in A^{u\ell}$. Since $A^{u\ell}$ is a down set this gives $\downarrow(\bigvee A) \subseteq A^{u\ell}$. Hence $A^{u\ell} = \downarrow(\bigvee A)$, as required. $\blacksquare$

2.33 Theorem. *Let P be an ordered set and define $\varphi : P \to \mathbf{DM}(P)$ by $\varphi(x) = \downarrow x$ for all $x \in P$.*

(i) $\mathbf{DM}(P)$ *is a completion of P via the map φ.*

(ii) φ *preserves all least upper bounds and greatest lower bounds which exist in P, that is, if $A \subseteq P$ and $\bigvee A$ exists in P, then $\varphi(\bigvee A) = \bigvee \varphi(A)$, and dually.*

Proof. (i) As we saw above, $\mathbf{DM}(P)$ is a complete lattice and φ is well-defined. It was noted in Lemma 1.21 that $x \leqslant y$ if and only if $\downarrow x \subseteq \downarrow y$; whence φ is an order-embedding.

(ii) Let $A \subseteq P$ and assume that $\bigvee A$ exists in P. We must show that $\varphi(\bigvee A) = \bigvee \varphi(A)$, that is, $\downarrow(\bigvee A) = \bigvee\{\downarrow a \mid a \in A\}$ in $\mathbf{DM}(P)$. We have $a \leqslant \bigvee A$ for all $a \in A$ and thus $\downarrow a \subseteq \downarrow(\bigvee A)$ for all $a \in A$. Hence $\downarrow(\bigvee A)$ is an upper bound for $\{\downarrow a \mid a \in A\}$ in $\mathbf{DM}(P)$. Let $B \in \mathbf{DM}(P)$ be an upper bound of $\{\downarrow a \mid a \in A\}$. Then $a \in \downarrow a \subseteq B$ for all $a \in A$, and so $A \subseteq B$. Consequently,

$$\downarrow(\bigvee A) = A^{u\ell} \subseteq B^{u\ell} = B;$$

the first equality holds by 2.32(iii), the inclusion holds since C is a closure operator, and the second equality holds since $B \in \mathbf{DM}(P)$.

Now assume that $\bigwedge A$ exists in P. We must show that $\varphi(\bigwedge A) = \bigwedge \varphi(A)$, that is,

$$\downarrow(\bigwedge A) = \bigwedge\{\downarrow a \mid a \in A\}.$$

Since $\mathbf{DM}(P)$ is a topped $\bigcap$-structure, by 2.17 we have

$$\bigwedge\{\downarrow a \mid a \in A\} = \bigcap\{\downarrow a \mid a \in A\} \quad \text{in} \quad \mathbf{DM}(P),$$

and hence 2.32(ii) yields the result. ∎

Before we can characterize the Dedekind–MacNeille completion, we need another definition and a lemma which is useful in its own right.

2.34 Definition. Let Q be an ordered set and let $P \subseteq Q$. Then P is called **join-dense** in Q if for every element $s \in Q$ there is a subset A of P such that $s = \bigvee_Q A$. The dual of join-dense is **meet-dense.**

To provide a compact notation, for all $s \in Q$ let

$$(\downarrow s)_P := \{y \in P \mid y \leqslant s\} \quad \text{and} \quad (\uparrow s)_P := \{y \in P \mid y \geqslant s\}.$$

2.35 Lemma. Let Q be an ordered set and let $P \subseteq Q$. The following are related by (i) $\Leftrightarrow$ (ii) $\Rightarrow$ (iii) in general and are equivalent if Q is a complete lattice:

(i) P is join-dense in Q;

(ii) $s = \bigvee_Q(\downarrow s)_P$ for all $s \in Q$;

(iii) for all $s, t \in Q$ with $t < s$ there exists $y \in P$ with $y \leqslant s$ and $y \not\leqslant t$.

Proof. This is left as an exercise. To show that (iii) implies (ii) when Q is a complete lattice, apply (iii) with $t := \bigvee_Q(\downarrow s)_P$. ∎

2.36 Theorem. Let P be an ordered set and let $\varphi : P \to \mathbf{DM}(P)$ be the order-embedding of P into its Dedekind–MacNeille completion given by $\varphi(x) = \downarrow x = \{y \in P \mid y \leqslant x\}$.

(i) $\varphi(P)$ is both join-dense and meet-dense in $\mathbf{DM}(P)$.

(ii) Let Q be an ordered set and let P be a subset of Q which is both join-dense and meet-dense in Q. Then there is an order-embedding ψ of Q into $\mathbf{DM}(P)$. Moreover, ψ agrees with φ on P, that is, $\psi(x) = \varphi(x)$ for all $x \in P$.

(iii) Let L be a complete lattice and let P be a subset of L which is both join-dense and meet-dense in L. Then $L \cong \mathbf{DM}(P)$ via an order-isomorphism which agrees with φ on P.

Proof. (i) Let $A \subseteq P$. We claim that in the complete lattice $\mathbf{DM}(P)$,

$$\bigwedge \{\, {\downarrow}x \mid x \in A^u \,\} = A^{u\ell} \quad \text{and} \quad \bigvee \{\, {\downarrow}a \mid a \in A \,\} = A^{u\ell}.$$

Indeed, for $B \subseteq P$, we have $B^\ell = \bigcap \{\, {\downarrow}x \mid x \in B \,\}$, since

$$y \in B^\ell \iff (\forall x \in B)\, y \leqslant x$$
$$\iff (\forall x \in B)\, y \in {\downarrow}x$$
$$\iff y \in \bigcap \{\, {\downarrow}x \mid x \in B \,\}.$$

With $B = A^u$ this yields the left-hand equality since meet is given in $\mathbf{DM}(L)$ by intersection. The proof of the right hand equality is only slightly more work. Let $a \in A$; then $a \in A^{u\ell}$ as $A \subseteq A^{u\ell}$ and hence ${\downarrow}a \subseteq A^{u\ell}$ as $A^{u\ell}$ is a down set. Hence $A^{u\ell}$ is an upper bound in $\mathbf{DM}(P)$ of $\{\, {\downarrow}a \mid a \in A \,\}$. If $B \in \mathbf{DM}(P)$ is an upper bound of $\{\, {\downarrow}a \mid a \in A \,\}$, then $a \in {\downarrow}a \subseteq B$ for all $a \in A$ and hence $A \subseteq B$. Thus $A^{u\ell} \subseteq B^{u\ell} = B$ (as $B \in \mathbf{DM}(P)$) and consequently $A^{u\ell}$ is the least upper bound of $\{\, {\downarrow}a \mid a \in A \,\}$ in $\mathbf{DM}(P)$.

Note that $\varphi(P) = \{\, \varphi(x) \mid x \in P \,\} = \{\, {\downarrow}x \mid x \in P \,\}$. If $A \in \mathbf{DM}(P)$, then $A^{u\ell} = A$ and hence, in $\mathbf{DM}(P)$, we have

$$\bigwedge \{\, {\downarrow}x \mid x \in A^u \,\} = A \quad \text{and} \quad \bigvee \{\, {\downarrow}a \mid a \in A \,\} = A.$$

Hence $\varphi(P)$ is both meet-dense and join-dense in $\mathbf{DM}(P)$.

(ii) Let $P \subseteq Q$ and assume that P is both join-dense and meet-dense in Q. We claim that the required order-embedding $\psi \colon Q \hookrightarrow \mathbf{DM}(P)$ may be defined by $\psi(s) := ({\downarrow}s)_P$ for all $s \in Q$. Firstly, we must show that ψ is well-defined, that is, $({\downarrow}s)_P \in \mathbf{DM}(P)$ for all $s \in Q$.

As P is join-dense in Q, 2.35 gives $s = \bigvee_Q ({\downarrow}s)_P$ for all $s \in Q$. (To avoid further subscripting, let us agree that if $A \subseteq P$, then A^u will denote the set of upper bounds of A *in* P, and similarly for A^ℓ. The corresponding constructs in Q will not be required.) Since $(({\downarrow}s)_P)^u$ is the set of all upper bounds (in P) of $({\downarrow}s)_P$ and since x is the least upper bound (in Q) of $({\downarrow}s)_P$, it follows that $(({\downarrow}s)_P)^u = ({\uparrow}s)_P$.

Dually, as P is meet-dense in Q, we obtain $((\uparrow s)_P)^\ell = (\downarrow s)_P$. Hence $((\downarrow s)_P)^{u\ell} = ((\uparrow s)_P)^\ell = (\downarrow s)_P$ and consequently $\psi(s) = (\downarrow s)_P \in \mathbf{DM}(P)$.

Clearly, $s \leqslant t$ implies $\psi(s) = (\downarrow s)_P \subseteq (\downarrow t)_P = \psi(t)$ for all $s, t \in Q$. Assume conversely that $\psi(s) \subseteq \psi(t)$, that is $(\downarrow s)_P \subseteq (\downarrow t)_P$ for some $s, t \in Q$. Then, since P is join-dense in Q and since $A \subseteq B$ implies $\bigvee A \leqslant \bigvee B$ (by 2.9), we obtain

$$s = \bigvee\nolimits_Q (\downarrow s)_P \leqslant \bigvee\nolimits_Q (\downarrow t)_P = t.$$

Thus $\psi \colon Q \hookrightarrow \mathbf{DM}(P)$. It is trivial that if $x \in P$ then $\psi(x) = \varphi(x)$.

(iii) Assume now that L is a complete lattice and P is a subset of L which is both join-dense and meet-dense in L. By (ii), the map given by $\psi(s) := (\downarrow s)_P$ for all $s \in L$ is an order-embedding of L into $\mathbf{DM}(P)$ such that $\psi(x) = \varphi(x)$ for all $x \in P$. Thus it remains to prove that ψ maps onto $\mathbf{DM}(P)$.

A slight modification of the proof of 2.32 shows that for all $A \subseteq P$ we have $A^{u\ell} = (\downarrow \bigvee_L A)_P$. (See Exercise 2.22.) Using this it is easy to see that ψ maps onto $\mathbf{DM}(P)$. Indeed, if $A \in \mathbf{DM}(P)$, then $A \subseteq P$ and $A^{u\ell} = A$. Hence if $s \in L$ is defined by $s := \bigvee_L A$, we have

$$\psi(s) = (\downarrow s)_P = A^{u\ell} = A,$$

whence ψ maps onto $\mathbf{DM}(P)$. $\blacksquare$

2.37 Remarks. The characterization of the Dedekind–MacNeille completion provided by this theorem may be used in several ways. For example, given an ordered set P, it may allow us to guess the structure of $\mathbf{DM}(P)$ without actually constructing the family of subsets of P which satisfy $A^{u\ell} = A$. Alternatively, given a complete lattice L it may enable us to find subsets P of L such that $L \cong \mathbf{DM}(P)$. (We see in Chapter 8 that for an important class of complete lattices L there is a *smallest* subset P with the property that $L \cong \mathbf{DM}(P)$.) These ideas are illustrated in the following examples.

2.38 Examples.

(1) It is not difficult to see that every real number $x \in \mathbb{R}$ satisfies $\bigvee_{\mathbb{R}} (\downarrow x)_{\mathbb{Q}} = x = \bigwedge_{\mathbb{R}} (\uparrow x)_{\mathbb{Q}}$ and hence $\mathbb{Q}$ is both join-dense and meet-dense in $\mathbb{R} \cup \{-\infty, \infty\}$. Consequently $\mathbb{R} \cup \{-\infty, \infty\}$ is (order-isomorphic to) the Dedekind–MacNeille completion of $\mathbb{Q}$.

(2) $\mathbf{DM}(\mathbb{N}) \cong \mathbb{N} \oplus \mathbf{1}$.

(3) For any set X, the complete lattice $\wp(X) \cong \mathbf{DM}(P)$ where

$$P = \{\, \{x\} \mid x \in X \,\} \cup \{\, X \smallsetminus \{x\} \mid x \in X \,\}.$$

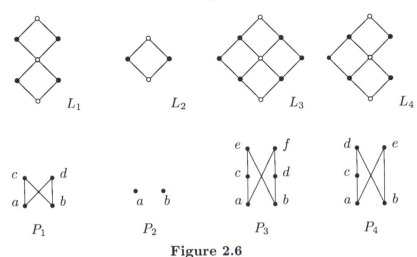

Figure 2.6

(4) The Dedekind–MacNeille completion of an n-element antichain (for $n \geqslant 1$) is order-isomorphic to the lattice $\mathbf{M}_n$ (see Figure 2.3).

(5) Each pair of diagrams in Figure 2.6 may be interpreted either as an ordered set P_i along with its Dedekind–MacNeille completion $L_i \cong \mathbf{DM}(P_i)$ or as a lattice L_i with a distinguished subset P_i such that $L_i \cong \mathbf{DM}(P_i)$. In each case the elements of P_i are shaded.

Exercises

Exercises from the text. Complete the proofs of 2.9, 2.10, 2.11, 2.12 (take care with the empty set!), 2.21 (the converse part). Prove the assertion that, if G is any group, $\mathrm{Sub}\,G$ is finite if and only if G is finite (see 2.8). Prove the assertion in 2.38(4).

2.1 Consider the diagram in Figure 2.7 of the ordered subset $P = \{1, 2, 3, 4, 5, 6, 7\}$ of $\langle \mathbb{N}_0; \leqslant \rangle$. Find the join and meet, where they exist, of each of the following subsets of P. Either specify the join or meet or indicate why it fails to exist.

(i) $\{3\}$, (ii) $\{4, 6\}$, (iii) $\{2, 3\}$, (iv) $\{2, 3, 6\}$, (v) $\{1, 5\}$.

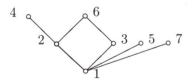

Figure 2.7

2.2 The ordered subset $Q = \{1, 2, 4, 5, 6, 12, 20, 30, 60\}$ of $\langle \mathbb{N}_0; \preccurlyeq \rangle$ is not a lattice. Draw a diagram of Q (it is 'cube-like') and find elements $a, b, c, d \in Q$ such that $a \vee b$ and $c \wedge d$ do not exist in Q.

2.3 Write out the duals of the following statements. (Each is true in all lattices, so its dual is also true in all lattices.)

(i) If z is an upper bound of $\{x, y\}$, then $x \vee y \leqslant z$.

(ii) $a \wedge b \leqslant a \leqslant a \vee b$ and $a \wedge b \leqslant b \leqslant a \vee b$.

(iii) If $a \prec c$ and $b \prec c$, with $a \neq b$, then $a \vee b = c$.

2.4 Consider the ordered set drawn in Figure 2.2. Use the technique illustrated in Example 2.4(3) to calculate the following elements or to explain why they fail to exist.

(i) $h \wedge i$ (iv) $(a \vee b) \vee c$ (vii) $a \vee (c \vee d)$
(ii) $a \vee c$ (v) $a \vee (b \vee c)$ (viii) $(a \vee c) \vee d$
(iii) $h \wedge j$ (vi) $c \vee d$ (ix) $\bigvee\{a, c, d\}$

2.5 Repeat Exercise 2.4 for the following elements of the ordered set shown in Figure 2.8.

(i) $\ell \wedge e$ (iv) $d \vee e$ (vii) $(j \wedge \ell) \wedge k$
(ii) $(\ell \wedge e) \wedge k$ (v) $c \vee (d \vee e)$ (viii) $j \wedge (\ell \wedge k)$
(iii) $c \vee e$ (vi) $\bigvee\{c, d, e\}$ (ix) $j \wedge k$

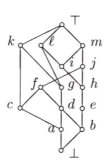

Figure 2.8

2.6 Give an example of an ordered set P in which there are three elements x, y, z such that

(a) $\{x, y, z\}$ is an antichain,

(b) $x \vee y$, $y \vee z$ and $z \vee x$ fail to exist,

(c) $\bigvee\{x, y, z\}$ exists.

(Of course, P will have more than three elements.)

2.7   A subset of $\mathbb{N}$ is called **cofinite** if $\mathbb{N} \setminus A$ is finite.

(i) Show that the collection $\mathcal{L}_1$ of cofinite subsets of $\mathbb{N}$ is a lattice of sets.

(ii) Show that the collection $\mathcal{L}_2$ of subsets of $\mathbb{N}$ which are either finite or cofinite is a lattice of sets.

(iii) Let $A_n := \mathbb{N} \setminus \{2, 4, \ldots, 2n\}$ be obtained from $\mathbb{N}$ by deleting the first n even natural numbers. Show that if $B \subseteq A_n$ for all $n \in \mathbb{N}$ then B is not cofinite and deduce that neither $\mathcal{L}_1$ nor $\mathcal{L}_2$ is complete.

2.8   Let L be a lattice. Prove that for all $a, b, c, d \in L$

(i) $a \leqslant b$ implies $a \vee c \leqslant b \vee c$ and $a \wedge c \leqslant b \wedge c$;

(ii) $a \leqslant b$ and $c \leqslant d$ imply $a \vee c \leqslant b \vee d$ and $a \wedge c \leqslant b \wedge d$.

(Note that (ii) says precisely that the binary operations $\vee \colon P^2 \to P$ and $\wedge \colon P^2 \to P$ are order-preserving.)

2.9   Let P be an ordered set and define $\varphi \colon P \to \mathcal{O}(P)$ by $\varphi(x) = \mathop{\downarrow} x$ for all $x \in P$.

(i) Show that φ is an order-embedding.

(ii) Prove that if P is a complete lattice, then there is a set X and a topped $\bigcap$-structure $\mathcal{L}$ on X such that $P \cong \mathcal{L}$. [Hint. Use Lemma 2.13 to show that the image of φ is a topped $\bigcap$-structure on P if P is complete.]

2.10   Prove that an order-isomorphism preserves all existing joins and meets. That is, if $\varphi \colon P \to Q$ is an order-isomorphism and $A \subseteq P$ such that $\bigvee A$ exists in P, then $\bigvee \varphi(A)$ exists in Q and $\bigvee \varphi(A) = \varphi(\bigvee A)$ (and dually for $\bigwedge A$).

2.11   Let $\langle L; \leqslant \rangle$ be a lattice. A non-empty subset J of L is called an **ideal** if

(a) $x, y \in J$ imply $x \vee y \in J$,

(b) if $x \in J$ and $y \in L$ with $y \leqslant x$, then $y \in J$.

(That is, J is closed under finite joins and closed under going down.) See Figure 2.9.

(i) Show that $\mathop{\downarrow} a$ is an ideal for all $a \in L$; it is known as the **principal ideal generated by** a.

shaded elements   shaded elements   shaded elements
an ideal   not an ideal   not an ideal

Figure 2.9

(ii) For each of the elements $a \in \mathbb{N}_0$ listed below, draw a diagram of the principal ideal $\downarrow a$ of the lattice $\langle \mathbb{N}_0; \preccurlyeq \rangle$:

$$a = 1, 2, 3, 6, 8, 12, 13, 16, 21, 24, 30, 36.$$

2.12 Let $\langle L; \leqslant \rangle$ be a lattice with a bottom element.

(i) Show that the family $\mathcal{I}(L)$ of ideals of L is a complete lattice when ordered by inclusion. [Hint. Prove that $\mathcal{I}(L)$ is a topped $\bigcap$–structure on L.]

(ii) Let $J_1 \subseteq J_2 \subseteq \ldots$ be a chain of ideals of L. Show that their union, $\bigcup \{ J_n \mid n \in \mathbb{N} \}$, is an ideal of L.

(iii) Show that every ideal of L is principal if and only if L satisfies (ACC). (In particular, every ideal of a finite lattice is principal.)

(iv) Give an example of a non-principal ideal in the chain $\mathbb{R}$.

2.13 The set $S = \{ (i,j) \in \mathbb{N} \times \mathbb{N} \mid i < j \}$ is given the order defined by

$$(i,j) \leqslant (i',j') \iff j \leqslant i' \text{ or } (i = i' \ \& \ j \leqslant j').$$

(i) Draw a diagram of S (as a subset of $\mathbb{N} \times \mathbb{N}$).

(ii) Show that S is a lattice having the property that, for any $n \geqslant 3$ and $a_1, \ldots, a_n \in S$, the element $a_1 \wedge a_2 \wedge \ldots \wedge a_n$ can be expressed as the meet of at most two of $a_1, \ldots, a_n$.

(iii) Show that the set of ideals of S, ordered by inclusion, is isomorphic to

$$\{ (i,j) \mid 1 \leqslant i < j \leqslant \infty \} \cup \{ (\infty, \infty) \}$$

with the order given by

$$(i,j) \leqslant (i',j') \iff j \leqslant i' \text{ or } (i = i' \ \& \ j \leqslant j').$$

2.14 Let L be a lattice and let $\varnothing \neq A \subseteq L$. Show that

$$(A] := \downarrow \{ a_1 \vee \ldots \vee a_n \mid n \in \mathbb{N}, \ a_1, \ldots, a_n \in A \}$$

is an ideal and moreover is the smallest ideal of L containing A. (Thus when L has a bottom, $(-]: \wp(L) \to \wp(L)$ is the closure operator corresponding to the topped $\bigcap$–structure $\mathcal{I}(L)$.)

2.15 Suppose that C is a closure operator on X and let $A \subseteq X$ and $A_i \subseteq X$ for each $i \in I$. Show that

(i) $C(\bigcup_{i \in I} A_i) \supseteq \bigcup\{ C(A_i) \mid i \in I \}$;

(ii) $C(A) \supseteq \bigcup\{ C(B) \mid B \subseteq A \text{ and } B \text{ is finite} \}$.

(These results are needed in Chapter 3.)

2.16 Draw $\mathrm{Sub}\,G$ and shade in the elements of $\mathcal{N}$-$\mathrm{Sub}\,G$ for each of the following groups: $\mathbf{S}_3$, $\mathbb{Z}_6$, $\mathbb{Z}_{12}$, $\mathbf{A}_4$, $\mathbf{D}_p$ (the symmetries of a regular p-gon) where p is an odd prime, the quaternion group.

2.17 Let H and K be finite groups such that $\gcd(|H|, |K|) = 1$. Show that $\mathrm{Sub}\,(H \times K) \cong \mathrm{Sub}\,H \times \mathrm{Sub}\,K$, where on the left we have the usual coordinatewise product of groups and on the right the coordinatewise product of ordered sets.

2.18 Prove that $P \times Q$ satisfies (ACC) if and only if both P and Q do.

2.19 Let P and Q be ordered sets of finite length. Prove that

$$\ell(P \times Q) = \ell(P) + \ell(Q).$$

2.20 Describe all lattices of length 2, proving that your list is complete.

2.21   (i) Show that $\langle \mathbb{N}_0; \preccurlyeq \rangle$ satisfies (DCC) and deduce that $\langle \mathbb{N}_0; \preccurlyeq \rangle$ is a complete lattice.

(ii) Prove directly that $\bigwedge S$ exists in $\langle \mathbb{N}_0; \preccurlyeq \rangle$ for every non-empty subset S of $\mathbb{N}_0$. [Hint. Let T be the set of prime divisors of the non-zero elements of S. Consider separately the cases where T is finite and T is infinite.]

2.22 Let L be a complete lattice, let P be a subset of L with the induced order and let $A \subseteq P$. Prove that $A^{u\ell} = (\downarrow\bigvee_L A)_P$. (Here $A^{u\ell}$ is calculated completely within P.)

2.23 Show that $\varnothing \in \mathbf{DM}(P)$ if and only if P has no bottom element.

2.24 Find the Dedekind–MacNeille completion of each of the ordered sets P_1, P_2, P_3, P_4 in Figure 2.6 and thereby verify that $L_i \cong \mathbf{DM}(P_i)$ in each case. [Use the definition of $\mathbf{DM}(P)$; do not appeal to Theorem 2.36.]

3

CPOs, Algebraic Lattices and Domains

This chapter looks in greater depth at $\bigcap$–structures. Aside from pure lattice theory, there is a double motivation for this study, from algebra and from computer science. The $\bigcap$–structures arising from algebra are usually topped and those in computer science are not, but there the dissimilarity ends. We therefore interweave the strands of theory relevant to the different areas of application.

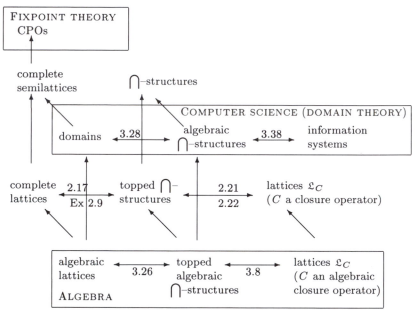

Figure 3.1

To help those who wish to select a particular strand, we give a diagram to show the interrelation between the various structures introduced in this chapter and in Chapter 2. A double arrow indicates a bijective correspondence and a single arrow an inclusion. The numbers refer to the subsections in which proofs of equivalence are given. The boxes indicate the main areas of application. The structures with 'algebraic' appended to their names have 'plenty of "finite" elements' and the various types of $\bigcap$–structures provide concrete examples or representations of the more abstract structures in the left-hand part of the diagram. Vertically we have a hierarchy of progressively weaker conditions on join and meet.

Directed joins and algebraic closure operators

All joins and meets exist in a complete lattice. In an ordered set, $\bigvee S$ cannot possibly exist unless $S^u \neq \varnothing$. On a topless ordered set P, such as one of our information-content examples, we might be tempted to take as a completeness condition the following: $\bigvee S$ exists whenever $S^u \neq \varnothing$. Thanks to Lemma 2.15, this forces $\bigwedge S$ to exist whenever $\varnothing \neq S \subseteq P$. In situations where we are concerned only with joins we might prefer a weaker condition with no implications for meets. This weakening will be the subject of the next section, which concerns CPOs. To proceed most directly to CPOs it is only necessary to master the definition of a directed set given in 3.1. In this section we motivate the introduction of directedness by looking at $\bigcap$-structures.

Consider the following assertions.

(1) Any group G is the union of its finitely generated subgroups.

(2) The graph of a map $f \colon \mathbb{N} \to \mathbb{N}$ is the union of the set of all graphs of partial maps $\sigma \leqslant f$ with $\operatorname{dom} \sigma$ finite (more informally, f can be built up from partial maps each specifying a finite amount of information about f).

Despite their very different ancestry, these examples have much in common. Each concerns an $\bigcap$-structure: $\operatorname{Sub} G$ in the first case and $(\mathbb{N} \relbar\joinrel\multimap \mathbb{N})$ (regarded, as in 2.19, as a family of sets) in the second. Each example concerns the approximation of a maximal, or total, element by 'finite' elements below it. Also, in each case we have a union, where we might have expected a lattice join. We surmise that our examples are drawn from a special class of $\bigcap$-structures. To confirm this we first seek conditions under which a join in an $\bigcap$-structure is given by set union. In general we have the unwieldy formula

$$\bigvee_{i \in I} A_i = \bigcap \{\, B \in \mathfrak{L} \mid \bigcup_{i \in I} A_i \subseteq B \,\},$$

from 2.17, for the join of a family $\{A_i\}_{i \in I}$ in an $\bigcap$-structure $\mathfrak{L}$. This may be replaced by $\bigvee_{i \in I} A_i = \bigcup_{i \in I} A_i$ precisely when $\bigcup_{i \in I} A_i \in \mathfrak{L}$. To see why this occurs in Examples (1) and (2) above, we need the definition of a directed set. It is convenient, for comparison, to couple this with the definition of a consistent set, to which we alluded in 1.6 and which plays an important part in the last section of this chapter.

3.1 Definitions. Let S be a non-empty subset of an ordered set P.

(i) S is said to be **directed** if, for every finite subset F of S, there exists $z \in S$ such that $z \in F^u$;

(ii) S is said to be **consistent** if, for every finite subset F of S, there exists $z \in P$ such that $z \in F^u$.

3.2 Remarks. Non-consistency arises only in ordered sets without $\top$. A directed set is, of course, consistent. An easy induction shows that S is directed if and only if, for every pair of elements $x, y \in S$, there exists $z \in S$ such that $z \in \{x, y\}^u$.

If $\mathcal{D} = \{A_i\}_{i \in I}$ is a directed set in $\wp(X)$ and $Y = \{y_1, \ldots, y_n\}$ is a finite subset of $\bigcup_{i \in I} A_i$ then there exists $A_k \in \mathcal{D}$ such that $Y \subseteq A_k$. Indeed, since $\mathcal{D}$ is directed, there exists $k \in I$ such that $y_j \in A_{i_j} \subseteq A_k$ for all j and consequently $Y \subseteq A_k$. This simple observation will be used repeatedly.

3.3 Examples.

(1) In any ordered set, any non-empty chain is directed and any subset with a greatest element is directed.

(2) The only directed subsets of an antichain are the singletons. More generally, in an ordered set with (ACC) (see 2.24 for the definition), a set is directed if and only if it has a greatest element.

(3) The finitely generated subgroups of a subgroup H of a group G form a directed subset of $\operatorname{Sub} G$. For any $f \in (\mathbb{N} \multimap \mathbb{N})$, the set

$$\{\, \sigma \in (\mathbb{N} \multimap \mathbb{N}) \mid \sigma \leqslant f \text{ and } \operatorname{dom} \sigma \text{ is finite} \,\}$$

is directed. Similarly, the set of finite subsets of Y is a directed subset of $\wp(X)$, for any $Y \subseteq X$. Can you suggest other examples of the same type?

The next lemma illustrates the role directed sets play in certain $\bigcap$-structures.

3.4 Lemma. *Let G be a group and $\mathcal{H} := \{H_i\}_{i \in I}$ be a directed subset of $\operatorname{Sub} G$. Then $H := \bigcup_{i \in I} H_i$ is a subgroup of G.*

Proof. Take $g_1, g_2 \in H$. It suffices to prove that $g_1 g_2^{-1} \in H$. For $j = 1, 2$, there exists H_{i_j} such that $g_j \in H_{i_j}$. Since $\mathcal{H}$ is directed, we can find $H_k \in \mathcal{H}$ so that $H_{i_1} \subseteq H_k$ and $H_{i_2} \subseteq H_k$. Then $g_1, g_2 \in H_k$, so $g_1 g_2^{-1} \in H_k$ because H_k is a subgroup. Hence $g_1 g_2^{-1} \in H$. ∎

3.5 Definition. An $\bigcap$-structure $\mathcal{L}$ is called **algebraic** if the union of any directed subfamily of $\mathcal{L}$ belongs to $\mathcal{L}$. Thus in an algebraic $\bigcap$-structure the join of any directed family is given by set union.

Lemma 3.4 shows that $\operatorname{Sub} G$ is an algebraic $\bigcap$-structure. Similarly, each of the $\bigcap$-structures listed in 2.23 can be shown to be algebraic.

Theorem 2.21 set up a correspondence between topped $\bigcap$-structures and closure operators. This specializes in a very satisfactory way to the algebraic case.

3.6 Definition. A closure operator C on a set X is called **algebraic** if, for all $A \subseteq X$,

$$C(A) = \bigcup \{ C(B) \mid B \subseteq A \text{ and } B \text{ is finite} \}.$$

It is easy to show that, for any closure operator C,

$$C(A) \supseteq \bigcup \{ C(B) \mid B \subseteq A \text{ and } B \text{ is finite} \}$$

(Exercise 2.15), so that to prove that a closure operator C is algebraic it is only necessary to prove the reverse inclusion.

3.7 Example. Recall from 2.23(1) that the closure operator corresponding to the $\bigcap$-structure $\text{Sub}\,G$ maps a subset A of G to the subgroup $\langle A \rangle$ generated by A. We claim that this closure operator is algebraic. This follows, via 3.8, from the fact that $\text{Sub}\,G$ is an algebraic $\bigcap$-structure, but the direct proof below is also instructive. By the remark in 3.6, it is sufficient to show that

$$\langle A \rangle \subseteq \bigcup \{ \langle B \rangle \mid B \subseteq A \text{ and } B \text{ is finite} \}.$$

Let $g \in \langle A \rangle$; then there exist $a_1, a_2, \ldots, a_n \in A$ such that $g = a_1' a_2' \ldots a_n'$, where $a_i' \in \{ a_i, a_i^{-1} \}$ for each i. Thus $g \in \langle \{ a_1, \ldots, a_n \} \rangle$, and this gives the required containment.

3.8 Theorem. *Let C be a closure operator on a set X and let $\mathfrak{L}_C$ be the associated $\bigcap$-structure. Then the following are equivalent:*

(i) *C is an algebraic closure operator;*

(ii) *for every directed family $\{ A_i \}_{i \in I}$ of subsets of X,*

$$C(\bigcup \{ A_i \mid i \in I \}) = \bigcup \{ C(A_i) \mid i \in I \};$$

(iii) *$\mathfrak{L}_C$ is an algebraic $\bigcap$-structure.*

Proof. Assume (i) holds and let $\{ A_i \}_{i \in I}$ be a directed family of subsets of X. First observe that if B is finite and $B \subseteq \bigcup_{i \in I} A_i$, then $B \subseteq A_k$ for some $k \in I$, by 3.2. Hence

$$C(\bigcup_{i \in I} A_i) = \bigcup \{ C(B) \mid B \subseteq \bigcup A_i \text{ and } B \text{ is finite} \} \qquad \text{(by (i))}$$

$$= \bigcup \{ C(B) \mid B \subseteq A_k \text{ for some } k \in I \text{ and } B \text{ is finite} \}$$

$$\text{(by the observation above)}$$

$$\subseteq \bigcup \{\, C(A_i) \mid i \in I \,\}.$$

The reverse inclusion is always valid (Exercise 2.15). Hence (i) $\Rightarrow$ (ii).

Since $\mathfrak{L}_C = \{\, C(A) \mid A \subseteq X \,\}$, it is trivial that (ii) implies (iii). Now assume (iii). Let $A \subseteq X$. The family $\mathcal{D} := \{\, C(B) \mid B \subseteq A$ and B is finite $\}$ is directed. Hence $\bigcup \mathcal{D} \in \mathfrak{L}_C$. Also $A \subseteq \bigcup \mathcal{D}$ since, for each $x \in A$, we have $x \in \{x\} \subseteq C(\{x\}) \subseteq \bigcup \mathcal{D}$. Hence

$$C(A) \subseteq C(\bigcup \mathcal{D})$$

$$= \bigcup \mathcal{D} \qquad\qquad (\text{since } \bigcup \mathcal{D} \in \mathfrak{L}_C)$$

$$= \bigcup \{\, C(B) \mid B \subseteq A \text{ and } B \text{ is finite} \,\}.$$

As previously noted, the reverse inclusion always holds. It follows that C is algebraic. ∎

CPOs

We have alluded several times to the idea of realizing an element in an ordered set as the join of a set of approximating elements below it. A set of approximations to a given element must be consistent; in the examples we have given, the approximations have in fact formed a directed set. Indeed, it is natural to regard $\bigvee S$ as the limit of S precisely when S is directed. We have so far only considered joins of directed sets in algebraic $\bigcap$–structures, where, as noted in 3.5, they coincide with unions. Arbitrary meets exist in an $\bigcap$–structure, yet it could be argued that they have no natural interpretation in a model for approximations. This leads us to study directed joins in a more general, meet-free, setting.

3.9 Definition. We say that an ordered set P is a **CPO** (an abbreviation for **complete partially ordered set**) if

(i) P has a bottom element, $\perp$,

(ii) $\bigvee D$ exists for each directed subset D of P.

Some authors do not require a CPO to have a bottom element and so omit (i), using the term **pointed CPO** when both (i) and (ii) hold. When we do not wish to assume $\perp$ is present in an ordered set P satisfying (ii) or do not want to regard $\perp$, even if it exists, as an integral part of the structure, we say P is a **pre-CPO**. Note that $P_\perp$ is a CPO whenever P is a pre-CPO.

We henceforth adopt the special notation $\bigsqcup D$ in place of $\bigvee D$ when the set D is directed.

3.10 Remarks. The simplest example of a directed set is a chain. Arguably, in the context of approximations, it would be adequate and less complicated to work with ascending chains rather than with directed sets. Indeed, in Chapter 4 this is exactly what we shall do. In fact it turns out that an ordered set with $\bot$ is a CPO if and only if it is **chain-complete** (that is, $\bigvee C$ exists for every non-empty chain). This is a non-trivial result. Exercise 3.5 seeks a proof in the countable case. The general case requires the machinery of ordinals.

3.11 Examples.

(1) Any complete lattice is a CPO.

(2) Let P be an antichain, or, more generally, any ordered set satisfying (ACC) (recall 2.24). Then P is a pre-CPO and $P_\bot$ is a CPO. In particular, $\mathbb{N}_\bot$ is a CPO. Indeed, this example is the primary motivation for the introduction of the lifting construction.

(3) Any algebraic $\bigcap$–structure is a CPO, with $\bigsqcup$ coinciding with $\bigcup$.

Many additional CPOs can be constructed by taking suitable subsets of CPOs or by combining CPOs in various ways.

3.12 Sub-CPOs.
Given a CPO P, we say that a subset Q of P is a **sub-CPO** if

(i) the bottom element of P belongs to Q,

(ii) whenever D is a directed subset of Q, the join $\bigsqcup_P D$ belongs to Q.

Condition (ii) is equivalent to

(ii)′ whenever D is a directed subset of Q, the join $\bigsqcup_Q D$ exists and coincides with $\bigsqcup_P D$.

In a pre-CPO, (i) is dropped. It is possible for a subset of a CPO P to be a CPO in its own right, without being a sub-CPO of P. For example, under the inclusion order,

$$\{\, S \subseteq \mathbb{N} \mid S \text{ is finite or } S = \mathbb{N} \,\}$$

is a CPO but is not a sub-CPO of $\wp(\mathbb{N})$: consider the join of the directed set $\{\, S \subseteq \mathbb{N} \setminus \{1\} \mid S \text{ finite} \,\}$.

We note for use in 4.14 that the intersection of a family of sub-CPOs of a CPO is itself a sub-CPO.

3.13 Sums and products of CPOs.
Let P and Q be CPOs. Then each of the following is a CPO:

$$P \oplus_\bot Q, \quad P \oplus_\vee Q, \quad P \times Q.$$

See 1.25 and 1.26 for the definitions. The first two assertions are elementary. Exercise 3.6 provides guidance on how to establish the third.

3.14 Continuous maps. In analysis, a function is continuous if it preserves limits. In a context in which a computation is modelled as the join (= limit) of a directed set, it is natural to consider a map as being continuous if it is compatible with the formation of directed joins; see also Exercise 3.11. Formally, we say that $\varphi \colon P \to Q$ (where P and Q are pre-CPOs) is **continuous** if, for every directed set D in P,

$$\varphi(\bigsqcup D) = \bigsqcup \varphi(D) \quad (:= \bigsqcup \{\, \varphi(x) \mid x \in D \,\}).$$

Note that since the empty set is not directed (by definition), a continuous map need not preserve bottoms. A map $\varphi \colon P \to Q$ such that $\varphi(\bot) = \bot$ is called **strict**. The natural structure-preserving maps for pre-CPOs are the continuous maps and for CPOs the strict continuous maps. The next lemma shows that every continuous map is order-preserving. Where the order represents 'is less defined than' or 'is a worse approximation than', an order-preserving map φ is one which is such that, the better the input x, the better the output $\varphi(x)$. Thus only maps which are order-preserving are likely to be of computational significance. For many applications the stronger property of continuity is the appropriate one.

3.15 Lemma. *Let P and Q be CPOs and φ be a map from P to Q.*

(i) *Suppose D is a directed subset of P and φ is order-preserving. Then $\varphi(D)$ is a directed subset of Q and $\bigsqcup \varphi(D) \leqslant \varphi(\bigsqcup D)$ and, in particular, $\bigsqcup_{n \geqslant 0} \varphi(x_n) \leqslant \varphi(\bigsqcup_{n \geqslant 0} x_n)$, for any ascending chain $x_0 \leqslant x_1 \leqslant x_2 \dots$.*

(ii) *If $\bigsqcup \varphi(D) \leqslant \varphi(\bigsqcup D)$ for every directed set D in P, then φ is order-preserving.*

Proof. The first part is an elementary application of the definitions. For the second, take x, y in P such that $x \leqslant y$. Then $D := \{x, y\}$ is directed, whence

$$\varphi(x) \leqslant \bigsqcup \varphi(D) \leqslant \varphi(\bigsqcup D) = \varphi(y). \qquad \blacksquare$$

3.16 Remarks.

(1) Not every order-preserving map between CPOs is continuous. Consider $\varphi \colon \wp(\mathbb{N}) \to \wp(\mathbb{N})$ defined by

$$\varphi(S) = \begin{cases} \varnothing & \text{if } S \text{ is finite,} \\ \mathbb{N} & \text{otherwise.} \end{cases}$$

The collection $\mathcal{D}$ of finite subsets of $\mathbb{N}$ is directed, with $\bigsqcup \varphi(\mathcal{D}) = \varnothing$ and $\varphi(\bigsqcup \mathcal{D}) = \mathbb{N}$.

(2) The continuity condition can be awkward to check. It is therefore useful to know when it is satisfied automatically. Using 3.3(2)

and 3.15(i), we see that an order-preserving map $\varphi \colon P \to Q$ is continuous whenever P satisfies (ACC), and in particular whenever P is a flat CPO (that is, a CPO $S_\perp$ for some set S).

The set of all continuous maps from the pre-CPO P to the pre-CPO Q, with the pointwise order, is denoted $[P \to Q]$. This function space construction provides an important way of building new CPOs. Note that $x \mapsto \perp$ acts as $\perp$ in $[P \to Q]$ whenever $\perp$ exists in Q.

3.17 Theorem. *Let P and Q be pre-CPOs. Then $[P \to Q]$ is a pre-CPO, and is a CPO whenever Q is a CPO.*

Proof. Let $E = \{\varphi_i\}_{i \in I}$ be a directed subset of $[P \to Q]$. For all $x \in P$, the subset $\{\varphi_i(x)\}_{i \in I}$ of Q is directed, since E is directed, and thus $\bigsqcup\{\varphi_i(x)\}_{i \in I}$ exists in the pre-CPO Q. Thus the pointwise join, $\varphi := \bigsqcup_{i \in I} \varphi_i$, of E is well defined and is order-preserving by 2.14. We establish that φ is continuous by showing that $\varphi(\bigsqcup D) \leqslant \bigsqcup \varphi(D)$ for an arbitrary directed subset D of P. Bracket-pushing, we have

$$\varphi(\bigsqcup D) = (\bigsqcup_{i \in I} \varphi_i)(\bigsqcup D)$$

$$= \bigsqcup_{i \in I}(\varphi_i(\bigsqcup D)) \qquad \text{(by the definition of } \bigsqcup \varphi_i\text{)}$$

$$= \bigsqcup_{i \in I}(\bigsqcup_{x \in D} \varphi_i(x)),$$

because each φ_i is continuous. But for each $x \in D$ and each $i \in I$,

$$\varphi_i(x) \leqslant \bigsqcup_{x \in D}(\bigsqcup_{i \in I} \varphi_i(x)).$$

Hence, for each $i \in I$,

$$\bigsqcup_{x \in D} \varphi_i(x) \leqslant \bigsqcup_{x \in D}(\bigsqcup_{i \in I} \varphi_i(x)).$$

(The left-hand side exists since, by 3.15(i), $\varphi(D)$ is directed.) So

$$\bigsqcup_{i \in I}(\bigsqcup_{x \in D}(\varphi_i(x))) \leqslant \bigsqcup_{x \in D}(\bigsqcup_{i \in I} \varphi_i(x)) = \bigsqcup_{x \in D} \varphi(x) = \bigsqcup \varphi(D). \qquad \blacksquare$$

We derive as a corollary a result which underpins many of the applications of the fixpoint theorems in Chapter 4. It brings maps and partial maps on subsets of $\mathbb{R}$ within the ambit of fixpoint theory. The equivalence in the corollary comes from 1.29.

3.18 Corollary. *For any set S the ordered set $(S \multimap S)$ (or equivalently $(S \to S_\perp)$) is order-isomorphic to the sub-CPO of $[S_\perp \to S_\perp]$ consisting of the strict maps (and so is itself a CPO).*

3.19 Complete semilattices. We introduced CPOs as a class of ordered sets in which suitable joins exist, but meets play no role, yet Lemma 2.15 shows that meets may sneak in by the back door. The situation is clarified by the following lemma, whose proof is left as an exercise. A CPO satisfying the equivalent conditions of the lemma is called a **complete semilattice**. Adjoining a top to such a CPO creates a complete lattice. It is not true that removal of the top from an arbitrary complete lattice leaves a complete semilattice; consider $\mathbb{N} \oplus \mathbf{1}$.

3.20 Lemma. *Let P be a CPO. Then the following are equivalent:*

(i) *P is **consistently complete**, that is, $\bigvee S$ exists for every consistent set S in P;*

(ii) *$\bigvee S$ exists whenever $S^u \neq \varnothing$;*

(iii) *$\bigwedge S$ exists whenever $S \neq \varnothing$;*

(iv) *$P \oplus \mathbf{1}$ is a complete lattice;*

(v) *$\downarrow a$ is a complete lattice for each $a \in P$.*

Finiteness, algebraic lattices and domains

We have referred several times to elements with a connotation of 'finiteness'. We now give the promised discussion of this concept. Two definitions are current, reflecting the confluence of ideas mentioned earlier: that of a **finite** element hails from computer science and that of a **compact** element from algebra (with the term cribbed from topology). Lemma 3.22 reconciles these definitions.

3.21 Definitions. Let P be a CPO and let $k \in P$. Then k is called **finite** (in P) if, for every directed set D in P,

$$k \leqslant \bigsqcup D \implies k \leqslant d \text{ for some } d \in D.$$

The set of finite elements of P is denoted $F(P)$.

Let L be a complete lattice and let $k \in L$. Then k is said to be **compact** if, for every subset S of L,

$$k \leqslant \bigvee S \implies k \leqslant \bigvee T \text{ for some finite subset } T \text{ of } S.$$

The set of compact elements of L is denoted $K(L)$.

3.22 Lemma. *Let L be a complete lattice. Then $F(L) = K(L)$. Further, $k_1 \vee k_2 \in F(L)$ whenever $k_1, k_2 \in F(L)$, and this remains true when L is merely a complete semilattice, provided $k_1 \vee k_2$ exists.*

Proof. Assume first that $k \in K(L)$ and that $k \leqslant \bigsqcup D$, where D is directed. Then there exists a finite subset F of D such that $k \leqslant \bigvee F$. Because D is directed, we can find $d \in D$ with $d \in F^u$. Then $k \leqslant d$, so $k \in F(L)$.

Conversely, assume that $k \in F(L)$ and that $k \leqslant \bigvee S$. The set

$$D = \{ \bigvee T \mid T \subseteq S \text{ and } T \text{ is finite} \}$$

is directed and $\bigsqcup D = \bigvee S$ (by Exercise 3.2). Applying the finiteness condition, we find a finite subset T of S with $k \leqslant \bigvee T$.

The second part is left as an exercise. ∎

3.23 Examples. Table 3.1 lists the finite elements in various complete lattices and CPOs. The assertions there can be checked directly or, in the case of the first four lattices, which are topped $\bigcap$–structures, obtained from Lemma 3.25. As a simple example of a non-finite element we have the top element of $\mathbb{N} \oplus \mathbf{1}$.

P	$F(P)$
$\wp(X)$ (X a set)	Finite subsets
$\mathcal{O}(P)$ (P an ordered set)	Sets $\downarrow\!F$ (F finite)
$\mathrm{Sub}\,G$ (G a group)	Finitely generated subgroups
$\mathrm{Sub}\,V$ (V a vector space)	Finite-dimensional subspaces
Σ^{**} (all binary strings)	Σ^* (finite strings)
$(X \multimap\!\!\rightarrow X)$ (partial maps on X)	Maps with finite domain
Ordered set P with (ACC)	All elements of P
$\mathbb{R}$	None

Table 3.1

In an $\bigcap$–structure, finiteness is closely bound up with being algebraic. Recall that a non-empty family $\mathfrak{L}$ of subsets of a set X is an algebraic $\bigcap$–structure if

(i) $\bigcap A_i \in \mathfrak{L}$ for any non-empty family $\{A_i\}_{i \in I}$ in $\mathfrak{L}$,

(ii) $\bigcup A_i \in \mathfrak{L}$ for any directed family $\{A_i\}_{i \in I}$ in $\mathfrak{L}$.

We now work towards a characterization of the ordered sets arising from algebraic $\bigcap$–structures, dealing with the topped case first.

3.24 Definition. A complete lattice L is said to be **algebraic** if, for each $a \in L$,

$$a = \bigvee \{ k \in K(L) \mid k \leqslant a \}.$$

The terminology will be justified by the next lemma which implies that many lattices arising in algebra are algebraic; see 3.29. In the following proofs, finiteness is more convenient to work with than compactness. Lemma 3.22 then allows results about algebraic lattices to be stated, as is traditional, in terms of compact elements.

3.25 Lemma. *Let C be an algebraic closure operator on X and $\mathfrak{L}_C$ the associated topped algebraic $\bigcap$–structure. Then $\mathfrak{L}_C$ is an algebraic lattice in which an element A is finite (equivalently, compact) if and only if $A = C(Y)$ for some finite set $Y \subseteq X$.*

Proof. We show that the finite elements are the closures of the finite sets. Then Definition 3.6 implies that $\mathfrak{L}_C$ is an algebraic lattice.

Let Y be a finite subset of X and let $A = C(Y)$. Take a directed set $\mathcal{D}$ in $\mathfrak{L}_C$ with $A \subseteq \bigsqcup \mathcal{D}$. Then, since $\bigsqcup$ coincides with $\bigcup$ in $\mathfrak{L}_C$,

$$Y \subseteq C(Y) = A \subseteq \bigsqcup \mathcal{D} = \bigcup \mathcal{D}.$$

As Y is finite and $\mathcal{D}$ directed, there exists $B \in \mathcal{D}$ such that $Y \subseteq B$. Then

$$A = C(Y) \subseteq C(B) = B,$$

so A is finite in $\mathfrak{L}_C$.

Conversely, assume that $A \in \mathfrak{L}_C$ is a finite element. Certainly

$$A = \bigsqcup \{ C(Y) \mid Y \subseteq A \text{ and } Y \text{ is finite} \}$$

(see 3.6). Invoke the finiteness of A in $\mathfrak{L}_C$ to find a finite set $Y \subseteq A$ such that $A \subseteq C(Y)$. The reverse inclusion holds since $Y \subseteq A$ implies $C(Y) \subseteq C(A) = A$. ∎

3.26 Theorem.

(i) *Let $\mathfrak{L}$ be a topped algebraic $\bigcap$–structure. Then $\mathfrak{L}$ is an algebraic lattice.*

(ii) *Let L be an algebraic lattice and for each $a \in L$ define $D_a := \{ k \in K(L) \mid k \leqslant a \}$. Then $\mathfrak{L} := \{ D_a \mid a \in L \}$ is a topped algebraic $\bigcap$–structure isomorphic to L.*

Proof. Part (i) follows from the preceding lemma and 3.8. For the converse, (ii), we leave the proof that $\mathfrak{L}$ is a topped $\bigcap$–structure as an exercise and prove here that the map $\varphi \colon a \mapsto D_a$ is an isomorphism of

L onto $\mathfrak{L}$ and that $\mathfrak{L}$ is algebraic. Because L is algebraic, $D_a \subseteq D_b$ in $\mathfrak{L}$ implies $a = \bigvee D_a \leqslant \bigvee D_b = b$ in L. The reverse implication holds always. Therefore φ is an order-isomorphism.

Take a directed subset $\mathcal{D} = \{ D_c \mid c \in C \}$ in $\mathfrak{L}$. As φ is an order-isomorphism, the indexing set C is a directed subset of L. Define $a = \bigsqcup C$. We claim that $\bigcup \mathcal{D} = D_a$ and so belongs to $\mathfrak{L}$. Indeed,

$$k \in D_a \Longleftrightarrow k \in K(L) = F(L) \text{ and } k \leqslant a = \bigsqcup C$$
$$\Longleftrightarrow k \in F(L) \text{ and } k \leqslant c \text{ for some } c \in C$$
$$\Longleftrightarrow k \in D_c \text{ for some } c \in C$$
$$\Longleftrightarrow k \in \bigcup \mathcal{D}.$$

Hence $\mathfrak{L}$ is closed under directed unions and so is algebraic. ∎

3.27 Domains. Our information-content examples mostly lack a top but are in all other respects very like algebraic lattices. If L is a complete semilattice (as defined in 3.19) then L is called an **algebraic semilattice** or a **domain** if, for all $a \in L$,

$$a = \bigsqcup \{ k \in F(L) \mid k \leqslant a \}.$$

(The join here is taken over a directed set, by 3.20 and 3.22.) The term domain is used widely in computer science, but not consistently (and certainly not directedly!); its meaning ranges from CPO through algebraic semilattice to algebraic semilattice with at most countably many finite elements. This last property is important in connection with computability questions which we shall not consider.

With the aid of Lemma 3.20 it is easily seen that adjoining a top to a domain gives an algebraic lattice and adjoining a top to an algebraic $\bigcap$–structure gives a topped algebraic $\bigcap$–structure. This observation yields a topless counterpart to Theorem 3.26, which in one direction provides a concrete realization of domains and in the other an order-theoretic characterization of algebraic $\bigcap$–structures.

3.28 Theorem.

(i) Let $\mathfrak{L}$ be an algebraic $\bigcap$–structure. Then $\mathfrak{L}$ is a domain.

(ii) Let L be a domain and for each $a \in L$ define $D_a := \{ k \in F(L) \mid k \leqslant a \}$. Then $\mathfrak{L} := \{ D_a \mid a \in L \}$ is an algebraic $\bigcap$–structure isomorphic to L.

3.29 Examples.

(1) The following are topped $\bigcap$–structures arising from algebraic closure operators and so are algebraic lattices (see 3.7 for a typical proof):

(i) $\wp(X)$, for any set X;

(ii) $\operatorname{Sub} G$, for any group G;

(iii) $\operatorname{Sub} V$, for any vector space V.

In addition, the chains **n**, for $n \geqslant 1$, and $\mathbb{N} \oplus \mathbf{1}$ are algebraic lattices.

(2) The following are domains:

(i) $(S \multimap\!\!\!\rightarrow S)$, for any $S \subseteq \mathbb{R}$;

(ii) Σ^{**} (all binary strings).

3.30 Remark: the role of continuity. Proposition 3.31 shows that a map φ from one domain, P, to another, Q, is continuous if and only if it is determined by finite approximations, that is, by its effect on the finite elements of P. Where domains are used as computational models, a finite element may be interpreted as an object conveying a finite amount of information. In this case the continuity condition asserts that to obtain a finite amount of information about $\varphi(x)$ it is only necessary to input a finite amount of information about x; this is exactly the import of (iii) in 3.31. Compare this with the requirement that φ be order-preserving, which may be informally stated as 'more information in implies more information out'.

3.31 Proposition. *Let P and Q be domains and $\varphi \colon P \to Q$ be order-preserving. Then the following are equivalent:*

(i) φ *is continuous;*

(ii) $\varphi(x) = \bigsqcup \{\, \varphi(k) \mid k \in F(P) \,\&\, k \leqslant x \,\}$ *for each $x \in P$;*

(iii) $D_{\varphi(x)} \subseteq \downarrow\!\varphi(D_x)$ *for all $x \in P$.*

Further, $[P \to Q]$ (the continuous maps from P to Q) is isomorphic to $\langle F(P) \to Q \rangle$ (the order-preserving maps from $F(P)$ to Q).

Proof. We have (i) $\Rightarrow$ (ii) because $D_x := \{ k \in F(P) \mid k \leqslant x \}$ is directed. Now assume (ii) holds and let $k' \in D_{\varphi(x)} = \bigsqcup \{\, \varphi(k) \mid k \in D_x \,\}$. Directedness implies $k' \leqslant \varphi(k)$ for some $k \in D_x$. Thus (iii) holds.

To prove (iii) $\Rightarrow$ (i), take $x := \bigsqcup D$. Let $k' \in D_{\varphi(x)}$. Thus, by (iii), $k' \leqslant \varphi(k)$ for some $k \in D_x$. Since $k \in F(P)$ and $k \leqslant x = \bigsqcup D$, we have $k \leqslant d$ for some $d \in D$. Thus $k' \leqslant \varphi(k) \leqslant \varphi(d) \leqslant \bigsqcup \varphi(D)$. Hence $\bigsqcup \varphi(D)$ is an upper bound of $D_{\varphi(x)}$ and consequently $\varphi(x) = \bigsqcup D_{\varphi(x)} \leqslant \bigsqcup \varphi(D)$.

Finally we note that the restriction map $\varphi \mapsto \varphi \!\restriction\! F(P)$ sets up an order-isomorphism from $[P \to Q]$ to $\langle F(P) \to Q \rangle$; it is onto because, for any $\psi \in \langle F(P) \to Q \rangle$, the map $x \mapsto \bigsqcup \{\, \psi(k) \mid k \in D_x \,\}$ is in $[P \to Q]$ and extends ψ. The details are left to the reader. ∎

Information systems

We already have two ways of presenting domains: abstractly, as algebraic semilattices, and, more concretely, as algebraic $\bigcap$–structures. A third way to arrive at domains is through information systems. This approach has a particularly intuitive appeal, because it capitalizes on the finiteness and information-content ideas which pervade the theory of computation. In this section we give a brief introduction to information systems and relate them to domains and algebraic $\bigcap$–structures. We contend that a judicious combination of the techniques and constructions derived from these alternative viewpoints gives an economical and intuitive approach to domain theory.

The starting point for the notion of an information system is the idea of identifying an object with a set of propositions true of it and adequate to define it. These propositions are to be thought of as tokens, each bearing a finite amount of information. Thus, if the objects to be described are the maps from $\mathbb{N}$ to $\mathbb{N}$, a suitable set of tokens would be $\mathbb{N} \times \mathbb{N}$, with the single token (m, n) true of f if and only if $f(m) = n$.

In the theory developed below, an information system has three constituents: a set A of tokens, a family Con of finite subsets of A (representing gobbets of consistent information) and a relation $\vdash$ of entailment (identifying implied, or superfluous, information). From an information system we build an $\bigcap$–structure $\mathfrak{L}$ on A so that each member of $\mathfrak{L}$ is a set of tokens whose finite subsets lie in Con and which contains all entailed tokens. The members of $\mathfrak{L}$, which are known as the elements of the information system, thus serve to represent the objects determined by consistent information. Further, $\mathfrak{L}$ is a domain and, conversely, every domain is associated with an information system in a canonical way.

3.32 Definition. An **information system** is a triple $\mathbf{A} = \langle A, Con, \vdash \rangle$ consisting of

(i) a set A of **tokens**;

(ii) a non-empty set Con of finite subsets of A (the finite **consistent** sets) which satisfy

   (IS)$_1$ $Y \in Con$ and $Z \subseteq Y$ implies $Z \in Con$,

   (IS)$_2$ $a \in A$ implies $\{a\} \in Con$;

(iii) $\vdash$ is a relation (**entailment**) between members of Con and members of A (formally $\vdash$ is a subset of $Con \times A$) satisfying

   (IS)$_3$ $Y \cup \{a\} \in Con$ whenever $Y \in Con, a \in A$ and $Y \vdash a$,

   (IS)$_4$ $Y \in Con$ and $a \in Y$ implies $Y \vdash a$,

(IS)$_5$ if $Y, Z \in Con$ and $a \in A$ are such that $Y \vdash b$ for all $b \in Z$ and $Z \vdash a$, then $Y \vdash a$.

Read $Y \vdash a$ as 'Y entails a' or 'a is deducible from Y'. An arbitrary subset $X \subseteq A$ is said to be **consistent** if every finite subset of X is in Con. We adopt the notation $Y \Subset A$ to mean that Y is a finite (possibly empty) subset of A.

3.33 Remarks. Axioms (IS)$_1$, (IS)$_3$ and (IS)$_4$ formalize commonsense features of consistency and entailment and (IS)$_2$ just says that every token contributes some information. The most mysterious axiom, (IS)$_5$, is a transitivity condition; it may be interpreted as saying that if from Y we can deduce enough information, Z, to deduce a, then we can deduce a from Y.

From (IS)$_2$ and (IS)$_1$ we deduce that $\varnothing$ (representing 'no information') is in Con. Further, (IS)$_1$ implies that a finite subset of A is consistent if and only if it belongs to Con.

The following rules are frequently used in proofs:

(a) if $Y \in Con$, $Z \subseteq Y$ and $Z \vdash a$, then $Y \vdash a$;

(b) if $Y \in Con$, Z is finite and $Y \vdash a$ for every $a \in Z$, then $Z \in Con$.

The first is entirely elementary. To prove the second, use induction on $|Z|$ to show $Y \cup Z \in Con$ and then appeal to (IS)$_1$.

3.34 Examples. Each of the triples $\mathbf{A} = \langle A, Con, \vdash \rangle$ in (1)–(5) below defines an information system, with the specified consistent sets. These examples show that the structure of an information system may be borne mainly by Con, by $\vdash$, or by the interaction between Con and $\vdash$.

(1) Take $A = \mathbb{N} \times \mathbb{N}$ and let

$\varnothing \in Con$ and $\{(m_1, n_1), \ldots, (m_k, n_k)\} \in Con$ if and only if $(m_i = m_j \Rightarrow n_i = n_j)$,

$Y \vdash (n, m)$ if and only if $(n, m) \in Y$.

The consistent sets are just the (graphs of) partial maps on $\mathbb{N}$.

(2) Let V be a vector space. Take $A = V$, Con to be all finite subsets of A and put $Y \vdash v$ if and only if $v \in \langle Y \rangle$, the subspace spanned by Y. All subsets of V are consistent.

(3) Take $A = \mathbb{N}$, let Con be all finite subsets of $\mathbb{N}$ and define

$\{n_1, \ldots, n_k\} \vdash n$ if and only if $n \leqslant n_i$ for some i.

All subsets of $\mathbb{N}$ are consistent.

(4) Take $A = \Sigma^*$ and let $\leqslant$ be the order on Σ^* defined in 1.6. Take

$Con = \{ Y \Subset \Sigma^* \mid \sigma, \tau \in Y \Rightarrow \sigma \leqslant \tau \text{ or } \tau \leqslant \sigma \},$

$Y \vdash \sigma$ if and only if $\sigma \leqslant \tau$ for some $\tau \in Y$.

A subset of A is consistent if and only if it is a subset of $\downarrow \sigma$ for some $\sigma \in \Sigma^{**}$.

(5) Recalling 1.6, take $A = \{ [\underline{x}, \overline{x}] \mid -\infty \leqslant \underline{x} \leqslant \overline{x} \leqslant \infty \}$ and define

$$\varnothing \in Con \text{ and } \{I_1, \ldots, I_k\} \in Con \text{ if and only if } I_1 \cap \cdots \cap I_k \neq \varnothing.$$

$$\{I_1, \ldots, I_k\} \vdash [\underline{x}, \overline{x}] \text{ if and only if } I_1 \cap \cdots \cap I_k \subseteq [\underline{x}, \overline{x}].$$

The non-empty consistent sets are the subsets of A with non-empty intersection.

(6) We may define a family of information systems $\perp_n$, for $n \geqslant 0$, as follows. Let $\perp_0$ be the unique information system with $\varnothing$ as its set of tokens. This has $\varnothing$ as its only consistent set and $\vdash$ such that there is no pair Y, a for which $Y \vdash a$. Now define inductively

$$a_0 = (0,0), a_1 = \{a_0, (1, a_0)\}, \ldots, a_{n+1} = \{a_n, (1, a_n)\}.$$

This creates a chain $a_0, a_1, \ldots$ of sets with $a_0 \in a_1$, $a_1 \in a_2 \ldots$; the process will be familiar to those who have seen the formal construction of the natural numbers. (This rather artificial notation is chosen to fit in with that used in 3.45(i). The pairing with 1 ensures that a new element is added at each stage.) For $n \geqslant 1$, take $\perp_n = \langle A_n, Con_n, \vdash_n \rangle$, where

$A_n = \{a_0, \ldots, a_{n-1}\}$,

Con_n consists of all subsets of A_n,

$Y \vdash a_i$ if and only if there exists $j \geqslant i$ such that $a_j \in Y$.

3.35 Elements. Each of the information systems above has a strong connection with some ordered set we have previously encountered. To make this precise, we need more definitions. Let $\mathbf{A} = \langle A, Con, \vdash \rangle$ be an information system. A set E of tokens is called an **element** of $\mathbf{A}$ if E is consistent and $\vdash$–closed, in the sense that $Y \in Con$, $Y \subseteq E$ and $Y \vdash a$ imply $a \in E$. The set of elements of $\mathbf{A}$ is denoted $|\mathbf{A}|$; in 3.37 we show that $|\mathbf{A}|$ is an algebraic $\bigcap$-structure.

For any consistent set X we define

$$\overline{X} := \{ a \in A \mid (\exists Y \subseteq X) \, Y \vdash a \};$$

this may be interpreted as the set of tokens deducible from X. When $X \in Con$, we have $\overline{X} = \{ a \in A \mid X \vdash a \}$. Lemma 3.36 shows that $\overline{X}$ is an element whenever X is consistent, and that every element is of this form. This lemma also characterizes elements in a way which reveals that, on those sets on which it is defined, the map $X \mapsto \overline{X}$ behaves very like an algebraic closure operator. Before stating the lemma we find the elements of the information systems in 3.34.

(1) $|\mathbf{A}| = (\mathbb{N} \multimap \mathbb{N})$, since all consistent sets are $\vdash$–closed.

(2) $|\mathbf{A}| = \operatorname{Sub} V$. This example typifies the way algebraic lattices come from information systems.

(3) $|\mathbf{A}| = \{\,\downarrow m \mid m \in \mathbb{N}\,\} \cup \{\mathbb{N}\}$. Here it may help to think of a 2-person game in which player two tries to find a member of $\mathbb{N} \cup \{\infty\}$ (corresponding to an element) fixed by player one, by making lower bound guesses (tokens).

(4) There is a one-to-one correspondence between the set Σ^{**} (finite or infinite binary strings) and the elements, under which a string is associated with its set of finite initial substrings. Under the inclusion order on $|\mathbf{A}|$, the elements finite in the CPO sense correspond to the finite strings and the maximal elements to the infinite strings.

(5) This example is rather similar to the last, with each element associated to an interval in $\mathbb{R} \cup \{-\infty, \infty\}$. Here the maximal elements correspond to the real numbers together with $-\infty$ and ∞.

(6) The elements of $\mathbf{1}_n$ are $\varnothing$ and the sets $\{a_0, \dots, a_k\}$ for $0 \leqslant k \leqslant n-1$. Ordered by inclusion, the elements form an n-element chain.

We prove one implication in the proof of the following lemma to illustrate how the rules in 3.33 and the axioms are employed and set the rest of the proof as an exercise.

3.36 Lemma. *Let $\mathbf{A} = \langle A, Con, \vdash \rangle$ be an information system and let $E \subseteq A$. Then the following are equivalent:*

(i) *E is consistent and $\vdash$–closed (that is, $E \in |\mathbf{A}|$);*

(ii) *$\{\overline{Y} \mid Y \in Con \text{ and } Y \Subset E\}$ is directed and*

$$E = \bigcup \{\overline{Y} \mid Y \in Con \text{ and } Y \Subset E\};$$

(iii) *$E = \overline{X}$ for some consistent set X.*

Proof of (iii) $\Rightarrow$ (i). Let $Z := \{x_1, \dots, x_k\} \Subset E$. For each i, there exists $Y_i \Subset X$ with $Y_i \in Con$ and $Y_i \vdash x_i$. Then $Y := Y_1 \cup \cdots \cup Y_k \Subset X$ and so $Y \in Con$, since X is consistent. By Rule (a) in 3.33, $Y \vdash x_i$ for each i. By Rule (b) in 3.33, $Z \in Con$. Hence E is consistent. To show E is $\vdash$–closed, assume $Z \vdash a$. The set Y above is such that $Y \vdash b$ for each $b \in Z$. By $(\mathrm{IS})_5$, $Y \vdash a$, so $a \in \overline{X} = E$. ∎

3.37 Information systems and algebraic $\bigcap$–structures. Take an information system $\mathbf{A} = \langle A, Con, \vdash \rangle$. We now prove our earlier claim that $|\mathbf{A}|$ is an algebraic $\bigcap$–structure, in other words a non-empty family of sets closed under intersections and directed unions of non-empty subfamilies. Since $|\mathbf{A}|$ contains $\overline{\varnothing}$, it is non-empty. It is routine to show that if $\{E_i\}_{i \in I}$ is a non-empty subfamily of $|\mathbf{A}|$ then $\bigcap_{i \in I} E_i$ is consistent and $\vdash$–closed, and so is in $|\mathbf{A}|$. Finally, assume $\mathcal{D} = \{E_i\}_{i \in I}$ is a directed set in $|\mathbf{A}|$ and let $E = \bigcup_{i \in I} E_i$. Take $Y \Subset E$. Because $\mathcal{D}$ is directed, $Y \Subset E_i$ for some i. Since E_i is consistent, we have $Y \in Con$. Assume also $Y \vdash a$. Then $a \in E_i$ since E_i is $\vdash$–closed, so $a \in E$. Therefore E is consistent and $\vdash$–closed. This completes the proof of the claim. It is an easy exercise to prove in addition that the finite elements of $|\mathbf{A}|$ are exactly the sets $\overline{Y}$ where $Y \in Con$.

In the other direction, take an algebraic $\bigcap$–structure $\mathfrak{L}$ and let $\mathbf{IS}(\mathfrak{L})$ be the information system $\langle A, Con, \vdash \rangle$ defined as follows:

(i) $A := \bigcup \mathfrak{L}$;

(ii) $Con := \{\, Y \mid (\exists U \in \mathfrak{L})\, Y \Subset U \,\}$;

(iii) $Y \vdash a$ if and only if $Y \in Con$, $a \in A$ and $a \in \bigcap \{\, U \in \mathfrak{L} \mid Y \Subset U \,\}$.

We outline the proof of the following theorem, leaving the reader to supply the details.

3.38 Theorem. *The maps* $\mathbf{A} \mapsto |\mathbf{A}|$ *and* $\mathfrak{L} \mapsto \mathbf{IS}(\mathfrak{L})$ *are mutually inverse and set up a bijective correspondence between the class of information systems and the class of algebraic* $\bigcap$*–structures.*

Proof. Given $\mathbf{A} = \langle A, Con, \vdash \rangle$, we claim that $\mathbf{A} = \mathbf{IS}(|\mathbf{A}|)$. We have

(i) $A = \bigcup |\mathbf{A}|$ (by $(\mathrm{IS})_2$);

(ii) if $Y \Subset A$, then $Y \in Con \Leftrightarrow (\exists E \in |\mathbf{A}|)\, Y \Subset E$ (for the forward implication note $\overline{Y} \in |\mathbf{A}|$ and for the reverse recall that any $E \in |\mathbf{A}|$ is consistent);

(iii) if $Y \in Con$ and $a \in A$, then $Y \vdash a \Leftrightarrow a \in \bigcap \{\, E \in |\mathbf{A}| \mid Y \Subset E \,\}$ (for the forward implication recall that any $E \in |\mathbf{A}|$ is $\vdash$–closed and for the reverse use the fact that $Y \subseteq \overline{Y} \in |\mathbf{A}|$).

Let $\mathfrak{L}$ be an algebraic $\bigcap$–structure. The formulae in 3.37 and 3.36 imply that, for an element E of $\mathbf{IS}(\mathfrak{L})$,

$$E = \bigcup \{\, \bigcap \{\, U \in \mathfrak{L} \mid U \supseteq Y \,\} \mid Y \Subset E \,\},$$

with the union taken over a directed set. Since $\mathfrak{L}$ is algebraic, $|\mathbf{IS}(\mathfrak{L})| \subseteq \mathfrak{L}$. Conversely, the definitions of consistency and entailment in $\mathbf{IS}(\mathfrak{L})$ imply that $\mathfrak{L} \subseteq |\mathbf{IS}(\mathfrak{L})|$. ∎

3.39 Information systems and domains. By combining 3.28 and 3.38 we obtain a bijective correspondence between information systems and domains. Given a domain D, the associated information system $\mathbf{IS}(D)$ has $F(D)$ (the finite elements of D) as its tokens, the finite sets consistent in the sense of 3.1 as the members of Con and $Y \vdash k$ if and only if $k \leqslant \bigvee Y$. Further $|\mathbf{IS}(D)|$ is order-isomorphic to D. The isomorphism is given by $U \mapsto \bigvee U$ and its inverse by $x \mapsto \{\, k \in F(D) \mid k \leqslant x \,\}$.

One cautionary remark needs to be made. Let D be the domain of elements of an information system $\mathbf{A}$. Then $|\mathbf{IS}(D)|$ is order-isomorphic to D but in general the set of tokens of $\mathbf{A}$ is quite different from the set of tokens of $\mathbf{IS}(D)$; indeed these sets of tokens may be of different cardinalities. Example 3.34(1) illustrates this well. The original set of tokens is $\mathbb{N} \times \mathbb{N}$, the domain of elements is $(\mathbb{N} {\multimap} \mathbb{N})$ and the token set of $\mathbf{IS}(D)$ is $\{\, \sigma \in (\mathbb{N} {\multimap} \mathbb{N}) \mid \operatorname{dom} \sigma \text{ is finite} \,\}$.

3.40 Technical remark. Take an algebraic $\bigcap$–structure $\mathfrak{L}$ on a set X. As Theorem 3.38 tells us, $|\mathbf{IS}(\mathfrak{L})|$ is the same family of sets as $\mathfrak{L}$. It is an $\bigcap$–structure on $\bigcup \mathfrak{L}$, which may be a proper subset of X. We define $\mathfrak{L}$ to be **full** if $\bigcup \mathfrak{L} = X$ or, in other words, if every point of the base set X belongs to a member of $\mathfrak{L}$. By construction $|\mathbf{A}|$ is full for any information system $\mathbf{A}$. Henceforth we work always with full $\bigcap$–structures and adopt the notation $(\mathfrak{L}, A)$ to indicate that $\mathfrak{L}$ is an $\bigcap$–structure with A as its base set.

3.41 Substructures and subsystems. Let $(\mathfrak{L}, A)$ and $(\mathfrak{K}, B)$ be algebraic $\bigcap$–structures. Then $\mathfrak{L}$ is called a **substructure** of $\mathfrak{K}$, and we write $\mathfrak{L} \sqsubseteq \mathfrak{K}$, if

(a) $A \subseteq B$,

(b) $\mathfrak{L} = \{\, U \cap A \mid U \in \mathfrak{K} \,\}$.

Note that we use the term substructure only in relation to $\bigcap$–structures which are algebraic. If $\mathfrak{L} \sqsubseteq \mathfrak{K}$ and $A = B$, then $\mathfrak{L} = \mathfrak{K}$. This useful property implies in particular that $\sqsubseteq$ is antisymmetric. It is obviously also reflexive and transitive. Therefore $\sqsubseteq$ defines a partial order on the set of substructures of any algebraic $\bigcap$–structure $\mathfrak{J}$. (Those who know a little set theory will appreciate that staying within some fixed $\mathfrak{J}$ is a device to ensure that we work with a set and not a proper class. This restriction causes no difficulties in practice, since it is usually possible to assume that all the $\bigcap$–structures involved in a given problem have their base sets lying in some fixed set X, so we may take $\mathfrak{J} = \wp(X)$.)

The ordering $\sqsubseteq$ of $\bigcap$–structures (or its information system equivalent, given below) is the key to solving domain equations (see 3.51 and 4.9). Exercise 3.29 elucidates how $\sqsubseteq$ works. It shows in particular that, if $\mathfrak{L} \sqsubseteq \mathfrak{K}$, then the map $S \mapsto \bigcap \{ T \in \mathfrak{K} \mid S \subseteq T \}$ is a continuous order-embedding of $\mathfrak{L}$ into $\mathfrak{K}$.

Let $\mathbf{A} = \langle A, Con_A, \vdash_A \rangle$ and $\mathbf{B} = \langle B, Con_B, \vdash_B \rangle$ be information systems. Then (an exercise) $|\mathbf{A}| \sqsubseteq |\mathbf{B}|$ if and only if

(i) $A \subseteq B$,

(ii) $Y \in Con_A \iff (Y \subseteq A \ \& \ Y \in Con_B)$,

(iii) $Y \vdash_A a \iff (Y \subseteq A \ \& \ a \in A \ \& \ Y \vdash_B a)$.

When these three conditions hold for $\mathbf{A}$ and $\mathbf{B}$, we say $\mathbf{A}$ is a **subsystem** of $\mathbf{B}$ and write $\mathbf{A} \trianglelefteq \mathbf{B}$. The relation $\mathbf{A} \trianglelefteq \mathbf{B}$ essentially says that $\mathbf{A}$ is less rich in information than $\mathbf{B}$, with each token of information from $\mathbf{A}$ providing a token in $\mathbf{B}$ with the same message. As a simple example we note that, for the information systems $\perp_n$ given in 3.34(6), we have

$$\perp_0 \trianglelefteq \perp_1 \trianglelefteq \ldots \trianglelefteq \perp_n \trianglelefteq \ldots.$$

We next look at the order properties of $\langle \mathrm{Sub}\,\mathfrak{J}; \sqsubseteq \rangle$, the family of substructures of $\mathfrak{J}$, and re-interpret these in terms of information systems. The proof of Theorem 3.42 is routine and we omit it.

3.42 Theorem. *Let $\mathfrak{J}$ be an algebraic $\bigcap$-structure. Then $\langle \mathrm{Sub}\,\mathfrak{J}, \sqsubseteq \rangle$ forms a domain.*

(i) *The bottom element is $\mathfrak{N} := \{\varnothing\}$, the unique algebraic $\bigcap$-structure based on $\varnothing$.*

(ii) *Let $\{(\mathfrak{L}_i, A_i)\}_{i \in I}$ be a non-empty family of substructures of $\mathfrak{J}$. Then*

$$\{ X \subseteq \bigcap_{i \in I} A_i \mid (\forall j \in I)(\exists Y_j \in \mathfrak{L}_j)\, X = Y_j \cap \bigcap_{i \in I} A_i \}$$

is an algebraic $\bigcap$-structure based on $\bigcap_{i \in I} A_i$ and equals $\bigwedge_{i \in I} \mathfrak{L}_i$.

(iii) *Let $\{(\mathfrak{L}_i, A_i)\}_{i \in I}$ be a family of substructures of $\mathfrak{J}$ directed with respect to $\sqsubseteq$. Then*

$$\{ X \subseteq \bigcup_{i \in I} A_i \mid (\forall j \in I)\, X \cap A_j \in \mathfrak{L}_j \}$$

is an algebraic $\bigcap$-structure on $\bigcup_{i \in I} A_i$ and equals $\bigsqcup_{i \in I} \mathfrak{L}_i$.

3.43 Proposition. *Let $\{\mathfrak{L}_i\}_{i \in I}$ be a non-empty family of substructures of the algebraic $\bigcap$-structure $\mathfrak{J}$ and let $\mathbf{IS}(\mathfrak{L}_i) = \langle A_i, Con_i, \vdash_i \rangle$.*

(i) $\mathbf{IS}(\bigwedge_{i \in I} \mathfrak{L}_i) = \langle \bigcap A_i, \bigcap Con_i, \bigcap \vdash_i \rangle$;

(ii) *if $\{(\mathfrak{L}_i, A_i)\}_{i \in I}$ is directed, $\mathbf{IS}(\bigsqcup_{i \in I} \mathfrak{L}_i) = \langle \bigcup A_i, \bigcup Con_i, \bigcup \vdash_i \rangle$.*

(Here the entailment relation $\vdash$ of $\langle A, Con, \vdash \rangle$ is taken to be the subset $\{(X, a) \mid X \vdash a\}$ of $Con \times A$.)

Proof. Apply Theorem 3.38. ∎

3.44 Constructions.   We have already noted that CPOs can be combined by forming sums, products and so on. It is far from clear that when we combine domains in such ways we remain within the class of domains. The three-way correspondence

$$\text{information system} \longleftrightarrow \text{algebraic } \bigcap\text{-structure} \longleftrightarrow \text{domain}$$

allows us to carry out constructions in whichever of these settings seems expedient and then use 3.37 and 3.38 to translate them into the formulation required. For the simpler constructions it is very convenient to work with algebraic $\bigcap$-structures, since it is quick and routine to check that the resulting structures are algebraic, so that the constructions do yield domains from domains. Proposition 3.45 collects a group of such results together and Example 3.46 illustrates the definitions in simple cases. We follow this with an example to show how the corresponding constructions for information systems can be read off, using Proposition 3.45. The other constructions are handled similarly.

At first sight, the definitions we give in 3.45 may look somewhat awkward. In (ii) the device of forming the product of the base sets A and B with $\{0\}$ and $\{1\}$ is necessary to force disjointness (see Exercise 1.21); from an information system viewpoint this is like colouring the tokens of two systems red and blue to keep track of which system they refer to. Similar tricks are employed in (i) and (iii). Also in (iii), the tokens in $\perp_{\mathfrak{L}}$ correspond to 'no information', so there is no harm in deleting them, so that $\perp_{\mathfrak{L}} = \varnothing$, and similarly for $\mathfrak{K}$. This is done in forming the coalesced sum.

3.45 Proposition.   *Let* $(\mathfrak{L}, A)$ *and* $(\mathfrak{K}, B)$ *be* $\bigcap$*-structures.*

(i) *Let* A_0 *be* $\{0\} \,\dot{\cup}\, A := \{(0,0)\} \cup (\{1\} \times A)$ *and define*

$$\mathfrak{L}_{\perp} = \{\, S \subseteq A_0 \mid S = \{(0,0)\} \cup (\{1\} \times T) \text{ for some } T \in \mathfrak{L} \,\} \cup \{\varnothing\}.$$

Then $\mathfrak{L}_{\perp}$ *is an* $\bigcap$*-structure on* A_0 *and is order-isomorphic to* $\mathbf{1} \oplus \mathfrak{L}$.

(ii) *Let* $A \,\dot{\cup}\, B = (\{0\} \times A) \cup (\{1\} \times B)$ *and define*

$$\mathfrak{L} \boxtimes \mathfrak{K} = \{\, (\{0\} \times S) \cup (\{1\} \times T) \mid S \in \mathfrak{L}, T \in \mathfrak{K} \,\}.$$

Then $\mathfrak{L} \boxtimes \mathfrak{K}$ *is an* $\bigcap$*-structure on* $A \,\dot{\cup}\, B$ *which is order-isomorphic to the ordered-set product* $\mathfrak{L} \times \mathfrak{K}$.

(iii) *Let* $A \,\dot{\cup}_{\vee}\, B$ *be* $(\{0\} \times (A \smallsetminus \perp_{\mathfrak{L}})) \cup (\{1\} \times (B \smallsetminus \perp_{\mathfrak{K}}))$ *and define*

$$\mathfrak{L} \boxplus_{\vee} \mathfrak{K} := \{\, \{0\} \times (S \smallsetminus \perp_{\mathfrak{L}}) \mid S \in \mathfrak{L} \,\} \cup \{\, \{1\} \times (T \smallsetminus \perp_{\mathfrak{K}}) \mid T \in \mathfrak{K} \,\} \cup \{\varnothing\}.$$

Then $\mathfrak{L} \boxplus_{\vee} \mathfrak{K}$ *is an* $\bigcap$*-structure on* $A \,\dot{\cup}_{\vee}\, B$ *which is order-isomorphic to the coalesced sum* $\mathfrak{L} \oplus_{\vee} \mathfrak{K}$.

(iv) $\mathfrak{L}\boxplus_\perp\mathfrak{K}:=\mathfrak{L}_\perp\boxplus_\vee\mathfrak{K}_\perp$ is an information system on $(A_0\,\dot\cup_\vee\,B_0)_0$ order-isomorphic to the separated sum $\mathfrak{L}\oplus_\perp\mathfrak{K}$.

Further, if $\mathfrak{L}$ and $\mathfrak{K}$ are algebraic, then so are $\mathfrak{L}_\perp$, $\mathfrak{L}\boxtimes\mathfrak{K}$, $\mathfrak{L}\boxplus_\vee\mathfrak{K}$ and $\mathfrak{L}\boxplus_\perp\mathfrak{K}$.

3.46 Example. We first illustrate 3.45(i). Take $\mathfrak{N} = \{\varnothing\}$. Then $\mathfrak{N}_\perp$ has base set $\{(0,0)\}$ and consists of the sets $\varnothing$ and $\{(0,0)\}$. Now repeat the process: $(\mathfrak{N}_\perp)_\perp$ has base set $\{(0,0),(1,(0,0))\}$ and consists of the sets $\varnothing$, $\{(0,0)\}$ and $\{(0,0),(1,(0,0))\}$. We have $\mathfrak{N} \sqsubseteq \mathfrak{N}_\perp \sqsubseteq (\mathfrak{N}_\perp)_\perp$, and so on. This chain of $\bigcap$–structures is just $|\underline{\perp}_0|, |\underline{\perp}_1|, |\underline{\perp}_2|, \ldots$.

Now consider $\mathfrak{N}_\perp\boxplus_\vee(\mathfrak{N}_\perp)_\perp$. This has base set

$$\{(0,(0,0)),(1,(0,0)),(1,(1,(0,0)))\}.$$

Its members are

$$\varnothing,\ \{(0,(0,0))\},\ \{(1,(0,0))\},\ \{(1,(0,0)),(1,(1,(0,0)))\};$$

as anticipated, $\mathfrak{N}_\perp\boxplus_\vee(\mathfrak{N}_\perp)_\perp$ with its inclusion order is isomorphic to $\mathbf{2}\oplus_\vee\mathbf{3}\cong\mathbf{1}\oplus(\mathbf{1}\,\dot\cup\,\mathbf{2})$.

3.47 Example. Take information systems $\mathbf{A} = \langle A, Con_A, \vdash_A\rangle$ and $\mathbf{B} = \langle B, Con_B, \vdash_B\rangle$. Their product is defined to be $\mathbf{IS}(|\mathbf{A}|\boxtimes|\mathbf{B}|)$, equal to $\langle C, Con, \vdash\rangle$, say. Proposition 3.45 tells us immediately that $C = A\,\dot\cup\,B$, as defined in 3.45(ii). To describe Con and $\vdash$, we write

$$X_0 := \{\,a \in A \mid (0,a) \in X\,\} \quad\text{and}\quad X_1 := \{\,b \in B \mid (1,b) \in X\,\},$$

for any $X \subseteq A\,\dot\cup\,B$. Then, again from 3.45,

$$Con = \{\,X \in C \mid X_0 \in Con_A\ \&\ X_1 \in Con_B\,\}$$

and $(X \vdash (0,a) \Leftrightarrow X_0 \vdash_A a)$ and $(X \vdash (1,b) \Leftrightarrow X_1 \vdash_B b)$.

3.48 Approximable mappings. In the applications of domain theory it is very important to know that the CPO of continuous maps from one domain to another is itself a domain. The search for an elementary route to this result has probably been largely responsible for the proliferation of alternative approaches to domains. We go via approximable mappings, which are to information systems what continuous maps are to domains (recall 1.16). An approximable mapping is not a map in the usual sense. It is an 'information-respecting' relation, to be thought of as a machine which from any finite set of information in one information system produces output in another.

Specifically, given information systems $\mathbf{A} = \langle A, Con_A, \vdash_A \rangle$ and $\mathbf{B} = \langle B, Con_B, \vdash_B \rangle$, an **approximable mapping** is a subset r of $Con_A \times B$, satisfying $(AM)_1$ and $(AM)_2$ below. We read $(Y, b) \in r$ as 'under r (the input) Y produces (the output) b'. Wherever possible it is best to replace $(Y, b) \in r$ by the more suggestive notation $r: Y \rightsquigarrow b$, or simply $Y \rightsquigarrow b$ when the name of the approximable mapping is not required. The axioms that must be satisfied are:

$(AM)_1$ $Y \rightsquigarrow b$ for all $b \in Z \Subset B$ implies

    (a) $Z \in Con_B$ and

    (b) $Z \vdash_B c \Rightarrow Y \rightsquigarrow c$;

$(AM)_2$ $Y \vdash_A a$ for all $a \in Y'$ and $Y' \rightsquigarrow b$ imply $Y \rightsquigarrow b$.

Here $\{\, X \mid X \rightsquigarrow b \text{ for some } b \in B \,\}$ should be thought as the inputs and $\{\, b \mid X \rightsquigarrow b \text{ for some } X \in Con_A \,\}$ as the information which is output.

The requirements on r are sensible. Part (a) of $(AM)_1$ says that putting consistent information into r gives consistent output, while (b) says that if the total output resulting from Y is enough to deduce c in B, then we get c itself as part of the output. Condition $(AM)_2$ ensures that adding extra consistent information to the input doesn't alter the output. Also, the conditions are exactly what is needed to make the next proof work.

The family of all approximable mappings is a family of subsets of $Con_A \times B$ and is readily seen to be an algebraic $\bigcap$-structure. Proposition 3.49 shows that this family gives the function space we require. Note the mix of $\bigcap$-structures and information systems.

3.49 Proposition. *Let $(\mathfrak{L}, A)$ and $(\mathfrak{K}, B)$ be algebraic $\bigcap$-structures. Then the family of approximable mappings from $\mathbf{IS}(\mathfrak{L})$ to $\mathbf{IS}(\mathfrak{K})$ is an algebraic $\bigcap$-structure which is order-isomorphic to $[\mathfrak{L} \to \mathfrak{K}]$. The isomorphism associates to $\varphi \in [\mathfrak{L} \to \mathfrak{K}]$ the approximable mapping s_φ given by $s_\varphi: Y \rightsquigarrow b$ if and only if $b \in \varphi(\overline{Y})$.*

Proof. A routine check confirms that s_φ is an approximable mapping whenever φ is a well-defined order-preserving map.

In the other direction, assume that r is an approximable map and define $|r|$ on $\mathfrak{L}$ by

$$|r|(U) := \{\, b \in B \mid (\exists Y \Subset U)\, r: Y \rightsquigarrow b \,\}.$$

To show that $|r|: \mathfrak{L} \to \mathfrak{K}$ it is enough by 3.38 to check that $|r|(U)$ is consistent and $\vdash$-closed for each $U \in \mathfrak{L}$. Let $Z := \{b_1, \ldots, b_n\} \Subset |r|(U)$. For each $i = 1, \ldots, n$, there exists $Y_i \Subset U$ such that $r: Y_i \rightsquigarrow b_i$. Then, by Rule (a) in 3.33 and $(AM)_2$, $r: Y \rightsquigarrow b_i$ for each i, where

$Y := Y_1 \cup \cdots \cup Y_n$. By $(\mathrm{AM})_1(\mathrm{a})$, Z is consistent in $\mathbf{IS}(\widehat{\mathfrak{K}})$, so $|r|(U)$ is consistent. To check $\vdash$–closure, assume also that $Z \vdash c$. By $(\mathrm{AM})_1(\mathrm{b})$, $r \colon Y \rightsquigarrow c$, whence $c \in |r|(U)$.

We claim that $|r| \in [\mathfrak{L} \to \widehat{\mathfrak{K}}]$. Let $\{U_i\}_{i \in I}$ be a directed family in $\mathfrak{L}$. The directedness implies that $Y \Subset \bigcup U_i$ if and only if $Y \Subset U_j$ for some j. It follows immediately that $|r|(\bigcup U_i) = \bigcup |r|(U_i)$, as required.

We now have maps $\Phi \colon \varphi \mapsto s_\varphi$ and $\Psi \colon r \mapsto |r|$. To prove that Φ and Ψ are mutually inverse bijections between $[\mathfrak{L} \to \widehat{\mathfrak{K}}]$ and the family of approximable mappings, we need $\varphi = |s_\varphi|$ and $r = s_{|r|}$. For $U \in \mathfrak{L}$,

$$
\begin{aligned}
\varphi(U) &= \varphi(\bigsqcup \{\, \overline{Y} \mid Y \Subset U \,\}) && \text{(by 3.36)} \\
&= \bigsqcup \{\, \varphi(\overline{Y}) \mid Y \Subset U \,\} && \text{(since } \varphi \text{ is continuous)} \\
&= \{\, b \in B \mid (\exists Y \Subset U)\, s_\varphi \colon Y \rightsquigarrow b \,\} && \text{(by the definition of } s_\varphi) \\
&= |s_\varphi|(U) && \text{(by the definition of } |\!-\!|).
\end{aligned}
$$

Let r be an approximable mapping. Then, for all $Y \in Con_A$ and $b \in B$,

$$
\begin{aligned}
s_{|r|} \colon Y \rightsquigarrow b &\iff b \in |r|(\overline{Y}) \\
&\iff (\exists Z \Subset Y)\, r \colon Z \rightsquigarrow b \\
&\iff r \colon Y \rightsquigarrow b && \text{(by } (\mathrm{AM})_2).
\end{aligned}
$$

The correspondence between $[\mathfrak{L} \to \widehat{\mathfrak{K}}]$ and the approximable mappings is an order-isomorphism provided Φ and Ψ are order-preserving. It follows from the definition that Ψ is order-preserving. Note also that $\varphi_1 \leqslant \varphi_2$ in $[\mathfrak{L} \to \widehat{\mathfrak{K}}]$ implies that $\varphi_1(\overline{Y}) \subseteq \varphi_2(\overline{Y})$ for all Y such that $Y \Subset U$ for some $U \in \mathfrak{L}$. But $s_{\varphi_i} \colon Y \rightsquigarrow b$ if and only if $b \in \varphi_i(\overline{Y})$ for $i = 1, 2$. It follows that $\varphi_1 \leqslant \varphi_2$ implies $s_{\varphi_1} \subseteq s_{\varphi_2}$, so Φ is order-preserving. ∎

Rephrasing our results about constructions in terms of domains we have the following theorem.

3.50 Theorem. *Let P and Q be domains. Then each of the following is also a domain:*

$$P_\perp,\ P \oplus_\perp Q,\ P \oplus_\vee Q,\ P \times Q,\ [P \to Q].$$

3.51 Domain constructors. The various constructions often need to be combined, in order to produce more complex domains. An n **to** m **domain constructor** maps an n–tuple of domains to an m–tuple of domains. Lifting, that is $P \mapsto P_\perp$, and $P \mapsto [[P \to P] \to P]$ are examples of **unary constructors** ($n = m = 1$). Taking $*$ equal to $\oplus_\perp, \oplus_\vee, \times$ or $\to$ in $(P, Q) \mapsto P * Q$, we obtain **binary constructors** ($n = 2, m = 1$).

We indicate in 3.52 how the search for mathematical models for programming languages leads to the problem of solving for P 'domain equations' $P \cong \mathbf{F}(P)$, where $\mathbf{F}$ is some domain constructor. Because of the way P is 'defined' in terms of itself, it is far from clear that a solution to this equation exists. Such apparent circularity is not an insuperable problem in simpler cases. It is easy to see that the equation $P \cong P_\perp$ is solved by taking $P = \mathbb{N} \oplus \mathbf{1}$ and it is almost as obvious that $P = \Sigma^{**}$ is a solution to $P \cong P \oplus_\perp P$. More complicated equations, for example $P \cong [P \to P]_\perp$, $P \cong Q \oplus_\perp [P \to P_\perp]$ or $P \cong Q \oplus_\perp (P \times P) \oplus_\perp [[P \to P] \to P]$ (where Q is fixed), look far less tractable.

A domain equation $P \cong \mathbf{F}(P)$ can be recast in the language of algebraic $\bigcap$–structures or of information systems. Then $\mathbf{F}$ may be treated as an operator on a CPO of algebraic $\bigcap$–structures ordered by $\sqsubseteq$ or on a CPO of information systems ordered by $\trianglelefteq$ (see 3.41). Results in Chapter 4 will show that the above equation is soluble if $\mathbf{F}$ is continuous, or, if we are willing to appeal to harder theory, just order-preserving.

Let $\mathfrak{X}$ be the family of substructures of some algebraic $\bigcap$–structure, ordered by $\sqsubseteq$. It is elementary that the lifting constructor preserves $\sqsubseteq$. To show that the binary constructors associated with sums, product and function space are order-preserving (as maps defined on $\mathfrak{X} \times \mathfrak{X}$) it is convenient to use the fact that the binary constructor $(\mathfrak{L}, \mathfrak{K}) \mapsto \mathfrak{L} * \mathfrak{K}$, where $\mathfrak{L}$ and $\mathfrak{K}$ range over $\mathfrak{X}$, is order-preserving if and only if each of the unary constructors $\mathfrak{L} \mapsto \mathfrak{L} * \mathfrak{K}$ (for fixed $\mathfrak{K}$) and $\mathfrak{K} \mapsto \mathfrak{L} * \mathfrak{K}$ (for fixed $\mathfrak{L}$) is order-preserving. To prove sufficiency, note that

$$
\begin{aligned}
(\mathfrak{L}_1, \mathfrak{K}_1) \sqsubseteq (\mathfrak{L}_2, \mathfrak{K}_2) &\implies \mathfrak{L}_1 \sqsubseteq \mathfrak{L}_2 \ \& \ \mathfrak{K}_1 \sqsubseteq \mathfrak{K}_2 \\
&\implies \mathfrak{L}_1 * \mathfrak{K}_1 \sqsubseteq \mathfrak{L}_2 * \mathfrak{K}_1 \ \& \ \mathfrak{L}_1 * \mathfrak{K}_1 \sqsubseteq \mathfrak{L}_2 * \mathfrak{K}_2 \\
&\implies \mathfrak{L}_1 * \mathfrak{K}_1 \sqsubseteq \mathfrak{L}_2 * \mathfrak{K}_2.
\end{aligned}
$$

The proof of necessity is equally easy. Checking this condition when $*$ is any of $\boxtimes$, $\boxplus_\vee$ or $\boxplus_\perp$ (as in 3.45) or $\to$ (as in 3.49) requires a clear head, but no ingenuity. It is recommended as an exercise to those wishing to become familiar with the definitions.

In fact, each of our constructors is continuous. To prove this we may first invoke Exercise 3.8, which reduces the problem to that of checking continuity of unary operators. By definition, a unary constructor $\mathbf{F}$ on

$\mathfrak{X}$ is continuous if and only if

$$\bigsqcup \mathbf{F}((\mathfrak{L}_i, A_i)) = \mathbf{F}(\bigsqcup (\mathfrak{L}_i, A_i))$$

for any directed family $\{(\mathfrak{L}_i, A_i)\}_{i \in I}$ in $\mathfrak{X}$. The remark following the definition of $\sqsubseteq$ in 3.41, together with 3.15, imply that this holds provided $\mathbf{F}$ is order-preserving and the $\bigcap$–structures $\bigsqcup \mathbf{F}((\mathfrak{L}_i, A_i))$ and $\mathbf{F}(\bigsqcup (\mathfrak{L}_i, A_i))$ have the same base set. We say that $\mathbf{F}$ is **continuous on base sets** if the latter condition is satisfied. In terms of information systems, this just means that every token of $\mathbf{IS}(\bigsqcup \mathfrak{L}_i)$ is a token drawn from $\mathbf{IS}(\mathfrak{L}_i)$ for some i. Checking the continuity of each of our domain constructors is now quite straightforward.

3.52 Denotational semantics and domain equations. We conclude this chapter with an informal account of the rudiments of denotational semantics, to indicate how domains and fixpoint theory underpin this approach to programming languages. Our discussion is perforce very brief and is aimed principally at readers with some experience of programming. References to full treatments of the subject can be found in the Appendix.

Suppose $\mathbf{L}$ is a programming language specified by formal syntactic rules. One way to analyze $\mathbf{L}$ is to construct a concrete mathematical model, $\mathbf{M}$, and a map $\mathcal{V}: \mathbf{L} \to \mathbf{M}$ (the valuation map), which to each object P in $\mathbf{L}$ (that is, a program constituent or a complete program) assigns $\mathcal{V}[\![P]\!]$ in $\mathbf{M}$, denoting the 'meaning' of P. For example, with $\mathbf{L}$ as the language Pascal, P might, for instance, be the multiplication operator $*$ and $\mathcal{V}[\![P]\!]$ the operation $\times$ of multiplication on $\mathbb{Z}$ (assuming that $\mathbf{M}$ is such that $\mathbb{Z} \subseteq \mathbf{M}$).

The abstract language $\mathbf{L}$ may regarded as being made up of various 'syntactic categories' (variables, commands, the expressions on which commands act, etc.), with complex program constructs defined in terms of simpler components. We seek to associate a 'semantic domain' to each syntactic category, to act as its concrete realization within $\mathbf{M}$. Starting from the most primitive of $\mathbf{L}$'s syntactic categories, we build up by stages the model $\mathbf{M}$, and the corresponding valuation map. This approach, which we illustrate below, is known as 'denotational semantics'. The model $\mathbf{M}$ enhances our understanding of $\mathbf{L}$ and enables us to reason about it. For example, we might seek to confirm that two programs in $\mathbf{L}$ have the same effect by showing they always have the same meaning in $\mathbf{M}$. In the other direction, the formal language may help us talk about $\mathbf{M}$. The study of propositional logic reveals a similar interplay between syntactic and semantic philosophies (see Chapter 7): the syntactic approach uses a formal language and the semantic approach uses truth

values. Predicate logic provides an even closer parallel. Denotational semantics is to programming languages what model theory is to predicate
logic.

Just as five year olds are taught to read from Ladybird books rather
than *War and Peace*, a study of semantics begins not with Pascal or
LISP, but with 'baby' languages, designed to exhibit particular programming features. High-level languages are built up from language
fragments in a modular fashion. We therefore start by looking at a very
simple imperative language, adequate only for doing arithmetic on $\mathbb{N}$.
This language has two primitive syntactic categories: *Val* (**basic values**) and *Id* (**identifiers**, which are our variables). The members of *Val*
are 'abstract versions' of the values our programs take. They fall into
two sets: the **numerals**, $Num = \underline{1}, \underline{2}, \ldots$, and the **Boolean values**,
$Bool = \{\mathbf{tt}, \mathbf{ff}\}$. The associated semantic domain is $V := \mathbb{N} \cup \{\mathbf{T}, \mathbf{F}\}$;
our notation for 'concrete' truth values is as in Chapter 7. The valuation
map takes each member of *Val* to the natural number or truth value it
'means'. Thus, $\mathcal{V}[\![\mathbf{tt}]\!] = \mathbf{T}$ and $\mathcal{V}[\![\underline{42}]\!] = 42$, etc.

Out of *Val* we build the **expressions**. These are of two types,
numeric and **Boolean**:

$$N ::= N \mid (N_1 + N_2) \mid (N_1 - N_2) \mid I$$
$$B ::= B \mid \neg B \mid (N_1 = N_2)$$

Here we have used what is called BNF notation, a shorthand way of stating which operations are admitted. In this case, each numeric expression
is either a numeral, the sum or difference of two numerals, or an identifier. A Boolean expression is either a Boolean value, the negation of
a Boolean expression, or the assertion that two numeric expressions are
equal. We could, of course, make our language more complex by adding
additional clauses to the prescription of the syntax, to represent further
operations. We use generic letters (subscripted or unsubscripted) to label
objects in the language: N for numeric expressions, E for expressions,
C for commands, and so on. We adopt the conventional notation and
use $\mathcal{E}$ and $\mathcal{C}$ to denote the restriction of $\mathcal{V}$ to *Exp* and *Cmd*.

The values assigned to identifiers are not constants. They change
as a program runs. A state, σ, records, as $\sigma(I)$, the current value of
each identifier, I. Formally σ is a map from *Id* to V. We then define
$\mathcal{E}[\![I]\!]\sigma = \sigma(I)$, for $\sigma \in St$, the set of states. This reflects our intention
that the meaning of an identifier at a given moment is the value currently
assigned to it. We take the semantic domain for *Exp* to be $(St \to V_\perp)$.

Language L	Valuation map $\xrightarrow{\;\;\nu\;\;}$	Model M
Syntactic categories		Semantic domains
Basic values, Val Numerals, Num Boolean values, $Bool$ Identifiers, Id		$V := \mathbb{N} \cup \{\mathsf{T},\mathsf{F}\}$ $\mathbb{N}$ $\{\mathsf{T},\mathsf{F}\}$ Id
Expressions, Exp Commands, Cmd	$\mathcal{E}$ $\mathcal{C}$	$(St \rightarrow V_{\perp})$ $(St_{\perp} \rightarrow St_{\perp})$

Table 3.2

This choice allows us to define $\mathcal{E}$ on compound expressions as follows:

$$\mathcal{E}\,[\![\,(N_1 + N_2)\,]\!]\,\sigma = \mathcal{E}\,[\![\,N_1\,]\!]\,\sigma + \mathcal{E}\,[\![\,N_2\,]\!]\,\sigma,$$

$$\mathcal{E}\,[\![\,(N_1 - N_2)\,]\!]\,\sigma = \begin{cases} \mathcal{E}\,[\![\,N_1\,]\!]\,\sigma - \mathcal{E}\,[\![\,N_2\,]\!]\,\sigma & \text{if this makes sense in } \mathbb{N}, \\ \perp & \text{otherwise,} \end{cases}$$

$$\mathcal{E}\,[\![\,\neg B\,]\!]\,\sigma = \mathsf{T} \iff \mathcal{E}\,[\![\,B\,]\!]\,\sigma = \mathsf{F},$$

$$\mathcal{E}\,[\![\,E_1 = E_2\,]\!]\,\sigma = \begin{cases} \mathsf{T} & \text{if } \mathcal{E}\,[\![\,E_1\,]\!]\,\sigma = \mathcal{E}\,[\![\,E_2\,]\!]\,\sigma, \\ \mathsf{F} & \text{otherwise.} \end{cases}$$

On the left hand sides, $+, -$ and $=$ are drawn from the formal language, while on the right they have their usual meanings in V. The set V is lifted so that $\perp$ can be used to model an 'error value'

Finally we need commands. These are 'state transformers': a command acts on a state and returns a new state. Their semantic domain is $(St_{\perp} \rightarrow St_{\perp})$. The lifting on the right allows for an error value. That on the left accommodates non-termination (looping): $\perp$ fed to a command is required to yield $\perp$. Our syntax for commands is

$$C ::= I := E \mid C_1; C_2 \mid \text{if } B \text{ then } C_1 \text{ else } C_2$$

This says that a command either assigns an expression to an identifier, or is two simpler commands in sequence (C_1 followed by C_2), or is an abstract version of if-then-else. The meanings of commands are determined as follows:

$$\mathcal{C}\,[\![\,I := E\,]\!]\,\sigma \text{ is the state which at } I' \neq I \text{ takes value } \sigma(I'),$$
$$\text{and at } I \text{ takes value } \mathcal{E}\,[\![\,E\,]\!]\,\sigma,$$

$$\mathcal{C}\,[\![\,C_1; C_2\,]\!]\,\sigma = (\mathcal{C}\,[\![\,C_2\,]\!] \circ \mathcal{C}\,[\![\,C_1\,]\!]\,)\sigma,$$

$$\mathcal{C}\,[\![\,\text{if } B \text{ then } C_1 \text{ else } C_2\,]\!]\,\sigma = cond\,(\mathcal{E}\,[\![\,B\,]\!]\,\sigma, \mathcal{C}\,[\![\,C_1\,]\!], \mathcal{C}\,[\![\,C_2\,]\!]\,)\sigma.$$

In the last definition, *cond* is the map from $(St \to \{\mathbf{T}, \mathbf{F}\}) \times St \times St$ to St given by

$$cond\,(P, f, g)\sigma = \begin{cases} f(\sigma) & \text{if } P(\sigma) = \mathbf{T}, \\ g(\sigma) & \text{otherwise.} \end{cases}$$

To add more elaborate commands, such as 'while B do C', which involve recursion, we need the machinery of Chapter 4; see 4.2 and 4.7.

Putting together the semantic domains for *Exp* and *Cmd* by forming the separated sum, we get our model for **L**. It is a domain, as defined in 3.27. The construction of **M** has the following features:

(i) the linguistic constructs on the primitive syntactic categories (+, =, etc.) are given global (that is, once-and-for-all) meanings,

and, fundamental to the philosophy of denotational semantics,

(ii) the meaning of a compound syntactic object is determined by the meanings of its syntactic subcomponents.

Setting up a semantic model for a functional language like LISP is much more difficult. Let us try to indicate, without setting up the syntax and semantics in full, what shape a model, **M**, for a language of this sort might have. We may take *Val* and *V* as before. We replace the previous categories *Exp* and *Cmd* by a single category of expressions. This encompasses our previous commands and expressions, except that the assignment clause $I := E$ is omitted and a mechanism for defining maps and procedures is substituted. The latter makes use of a new syntactic category of 'declarations'. Values of compound programs are to be determined from the values of their parts, so we must allow **M** to contain maps. Maps may well need to act on other maps and we may require procedures to act on maps or even other procedures. We therefore do not want to tie ourselves down by specifying the 'type' of the argument of a map or procedure. This suggests we should take as the semantic domain for expressions the domain $(\mathbf{M} \to \mathbf{M}_\perp)$, so that the meanings of expressions may be maps defined on any part of **M**; recall 1.29. When all this is fleshed out and made precise (no small undertaking) it leads to the conclusion that a model **M** for a functional language might need to satisfy $\mathbf{M} \cong V \oplus_\perp (\mathbf{M} \to \mathbf{M}_\perp)$. This is patently impossible on cardinality grounds, by Cantor's Theorem (see [8]). The cardinality problem disappears if instead of *all* maps from **M** to $\mathbf{M}_\perp$ we take $[\mathbf{M} \to \mathbf{M}_\perp]$. Since it is the continuous maps that are the computationally significant ones (recall 3.30), this is a sensible modification in any case. We are now faced with the problem of whether there is a domain **M** satisfying

$$\mathbf{M} \cong V \oplus_\perp [\mathbf{M} \to \mathbf{M}_\perp].$$

The fixpoint theory in the next chapter is just what is needed to show that this equation is soluble, along with many others of a similar kind. The existence of domains to serve as models for functional languages is thus ensured.

Exercises

Exercises from the text. Prove the assertions about sums in 3.13. Prove 3.15(i) and Corollary 3.18. Prove Lemma 3.20. (Show (ii)–(v) are equivalent for any ordered set P, that (i) implies (ii) and that (ii) implies (i). [Hint. Use Lemma 2.15 and Exercise 3.3].) Prove the assertions about $F(P)$ in Table 3.1 in case $P = \Sigma^{**}$, $P = (X \multimap X)$ or P satisfies (ACC).

Complete the proofs of 3.22, 3.26(ii) and 3.31. Verify that the examples in 3.34 are indeed information systems, with elements as specified in 3.35. Prove Rules (a) and (b) in 3.33. Complete the proof of Lemma 3.36 by proving the implications (i) $\Rightarrow$ (ii) $\Rightarrow$ (iii). Prove the claim in 3.37 that $F(|\mathbf{A}|) = \{ \overline{X} \mid X \in Con \}$ for any information system $\mathbf{A}$. Complete the proof of Theorem 3.38. Prove the claim in 3.41 that for information systems $\mathbf{A}$ and $\mathbf{B}$ we have $|\mathbf{A}| \sqsubseteq |\mathbf{B}|$ if and only if conditions (i)–(iii) given there hold. Prove Theorem 3.42.

3.1   Determine in which of the following cases D is (a) a consistent subset of P, (b) a directed subset of P:

   (i) $P = \wp(\mathbb{N})$, $D = \{ X \subseteq \mathbb{N} \mid X \text{ is finite} \}$;

   (ii) $P = \wp(\mathbb{N})$, $D = \{ X \subseteq \mathbb{N} \mid \mathbb{N} \smallsetminus X \text{ is finite} \}$;

   (iii) $P = (\mathbb{N} \multimap \mathbb{N})$, $D = \{ \pi \in P \mid \pi(n) \leqslant 2 \text{ for all } n \in \operatorname{dom} \pi \}$;

   (iv) $P = \mathbb{N}^2$, $D = \{ (0, n) \mid n \in \mathbb{N} \} \cup \{ (n, 0) \mid n \in \mathbb{N} \}$;

   (v) $P = (\mathbb{N}_0 \to \mathbb{N}_0)$, $D = \{ f_i \}_{i \geqslant 1}$, where $f_i \colon \mathbb{N}_0 \to \mathbb{N}_0$ is defined by $f_i(i) = 1$ and $f_i(j) = 0$ for $j \neq i$;

   (vi) $P = \langle \mathbb{N}; \preccurlyeq \rangle$, $D = \{ 2^a 3^b \mid a, b \in \mathbb{N} \text{ and } a + b \text{ is prime} \}$.

3.2   Let P be a pre-CPO. Assume that S is a non-empty subset of P such that $\bigvee F$ exists for every finite non-empty subset F of S. Prove that $\bigvee S$ exists by showing that

$$\bigvee S = \bigsqcup \{ \bigvee F \mid \varnothing \neq F \Subset S \}$$

3.3   Let P be a pre-CPO and let $A_i \subseteq P$ for all $i \in I$. Suppose that A_i is a directed subset of P for all $i \in I$ and that $\{ A_i \}_{i \in I}$ is a directed family of subsets of P. Show that both $\bigcup_{i \in I} A_i$ and $\{\bigsqcup A_i\}_{i \in I}$ are directed subsets of P and that

$$\bigsqcup (\bigcup_{i \in I} A_i) = \bigsqcup_{i \in I} (\bigsqcup A_i).$$

3.4 Which of the following are (pre)-CPOs?

(i) $\{0, 1, 2\}^*$ (the set of finite strings of zeros, ones and twos) with order defined as in 1.6);

(ii) the set of all finite strings of zeros and ones with order defined as in Exercise 1.8;

(iii) the family of all countable (including finite) subsets of $\mathbb{R}$, ordered by inclusion;

(iv) $\{1/n \mid n = 1, 2, \ldots\} \cup \{0\}$;

(v) the chain $\mathbb{Q} \cap [0, 1]$;

(vi) $\{1 - 1/2^n - 1/5^m \mid n, m = 1, 2, \ldots\}$.

(In (iv)–(vi) the order is that induced from the chain $\mathbb{R}$.)

3.5 Let P be a countable ordered set such that $\bigsqcup C \ (= \bigvee C)$ exists for every chain C in P and let $D = \{x_0, x_1, x_2, \ldots\}$ be a directed subset of P. For each finite subset F of D let u_F be an upper bound for F in D. Define sets D_i as follows:

$$D_0 = \{x_0\}, \qquad D_{i+1} = D_i \cup \{y_{i+1}, u_{D_i \cup \{y_{i+1}\}}\},$$

where y_{i+1} is the element x_n in $D \smallsetminus D_i$ with subscript n chosen as small as possible. Prove that

(i) for each i, the set D_i is directed and has at least i elements;

(ii) the sets D_i form a chain;

(iii) $\{\bigvee D_i\}_{i \geqslant 1}$ is a chain in P and its join is $\bigsqcup D$.

3.6 Let P_1 and P_2 be CPOs. Let $D \subseteq P_1 \times P_2$ be a directed set. Define D_1 and D_2 by

$$D_1 := \{x_1 \in P_1 \mid (\exists x_2 \in P_2)\, (x_1, x_2) \in D\},$$
$$D_2 := \{x_2 \in P_2 \mid (\exists x_1 \in P_1)\, (x_1, x_2) \in D\}.$$

Show that D_1 and D_2 are directed and that $\bigsqcup D = (\bigsqcup D_1, \bigsqcup D_2)$. Deduce that $P_1 \times P_2$ is a CPO.

3.7 Let P_1 and P_2 be CPOs. Define the projections π_1 and π_2 by $\pi_i(x_1, x_2) = x_i \quad (i = 1, 2)$.

(i) Prove that $\pi_1 \colon P_1 \times P_2 \to P_1$ and $\pi_2 \colon P_1 \times P_2 \to P_2$ are continuous.

(ii) Prove that a map $\varphi \colon Q \to P_1 \times P_2$, where Q is another CPO, is continuous if and only if both $\pi_1 \circ \varphi$ and $\pi_2 \circ \varphi$ are continuous.

3.8 Let P_1, P_2 and Q be CPOs and let $\varphi \colon P_1 \times P_2 \to Q$ be a map.

(i) Define φ^x and φ_y by

$$\varphi^x(y) = \varphi(x, y)\,(y \in P_2) \quad \text{and} \quad \varphi_y(x) = \varphi(x, y)\,(x \in P_1).$$

Prove that φ is continuous if and only if φ^x and φ_y are continuous for all x, y (that is, φ is continuous if and only if it is continuous in each variable separately).

(ii) Let φ be continuous and define $\Phi \colon P_1 \to [P_2 \to Q]$ by

$$(\Phi(x))(y) = \varphi(x, y) \quad (x \in P_1, y \in P_2).$$

Prove that Φ is well-defined and continuous and deduce that $[(P_1 \times P_2) \to Q] \cong [P_1 \to [P_2 \to Q]]$.

3.9 Let C be a closure operator on a set X. Show that C is algebraic if and only if it is continuous as a map from $\wp(X)$ to $\wp(X)$.

3.10 Which of the following maps $\varphi \colon \wp(\mathbb{N})^2 \to \wp(\mathbb{N})$ are continuous?

(i) $\varphi(S, T) = S \cup T$;

(ii) $\varphi(S, T) = S \cap T$;

(iii) $\varphi(S, T) = S \smallsetminus T$.

[Hint. Use Exercise 3.8.]

3.11 Let P be a CPO. Let $\mathcal{F}$ be the family of sets $U \in \mathcal{O}(P)$ such that D directed and $D \subseteq U$ imply $\bigsqcup D \in U$.

(i) Identify $\mathcal{F}$ when P is

(a) $\mathbb{N} \oplus \mathbf{1}$,   (b) $\overline{\mathbf{2}}_\perp$,   (c) $\wp(\mathbb{N})$,   (d) $(\mathbb{N}\!\multimap\!\!\to\mathbb{N})$.

(ii) Show that $\mathcal{F}$ contains $\varnothing$ and P and is closed under arbitrary intersections and finite unions (and so is the family of closed sets for a topology $\mathcal{T}$ on P (the **Scott topology**)).

(iii) Let P and Q be CPOs and topologize each as above. Prove that a map $\varphi \colon P \to Q$ is topologically continuous if and only if it is continuous in the CPO sense. [Hint. Recall that φ is topologically continuous if and only if $\varphi^{-1}(V)$ is closed in P whenever V is closed in Q.]

3.12 Give an example of a CPO P and elements $x \notin F(P), k \in F(P)$ such that $x < k$.

3.13 Identify the finite elements in those ordered sets in Exercise 3.4 which are CPOs.

3.14 Let L be an algebraic lattice and K a non-empty subset of L such that $\bigvee_L S$ and $\bigwedge_L S$ belong to K for every non-empty subset S of K. Show that K is an algebraic lattice. [Hint. Represent L as a topped algebraic $\bigcap$-structure on some set X and show that K is also a topped algebraic $\bigcap$-structure on some subset Y of X.]

3.15 Let R be a commutative ring with identity, 1, and let $\mathfrak{L}$ be the $\bigcap$-structure of all ideals of R. Prove that $R \in K(\mathfrak{L})$.

3.16 Let $P = \{\,(a,b) \mid a, b \subseteq \mathbb{N}, a \cap b = \varnothing\,\}$, ordered as a subset of $\wp(\mathbb{N}) \times \wp(\mathbb{N})$. When are two elements (a,b) and (c,d) consistent? Which elements of P are maximal? Which elements are finite? Prove that P is a domain.

Prove that P is isomorphic to the set of all maps from $\mathbb{N}$ to $\mathbf{2}_\perp$ with the pointwise order.

3.17 Show that if D is a domain then $D \oplus 1$ is an algebraic lattice. [Hint. First consider the particular case where D itself is an algebraic lattice.]

3.18 A non-empty ordered set S is called a **join semilattice** if $x \vee y$ exists in S for all $x, y \in S$. As in the lattice case, a non-empty subset J of S is an **ideal** of S if J is closed under finite joins and going down. (See Exercise 2.11.)

  (i) Show that the set $\mathcal{I}(S)$ of ideals of a join semilattice with $\perp$ is a topped algebraic $\bigcap$-structure on S and that $K(\mathcal{I}(S)) \cong S$.

  (ii) By Lemma 3.22, $K(L)$ is a join semilattice with $\perp$ for any complete lattice L. Prove that the following are equivalent:

  (a) L is an algebraic lattice;

  (b) $L \cong \mathcal{I}(K(L))$;

  (c) $L \cong \mathcal{I}(S)$ where S is some join semilattice with $\perp$.

[Hint. Show that J is an ideal of $K(L)$ if and only if J is of the form D_a for some $a \in L$, then appeal to Theorem 3.26.]

3.19 Let Q be a non-empty ordered set. A non-empty subset J of Q is called an **ideal** of Q if it is a directed down-set. The set of all ideals of Q is denoted by $\mathcal{I}(Q)$.

(i) Show that if Q is a join semilattice then this concept of ideal agrees with the one introduced in Exercise 3.18.

(ii) Give an example of an ordered set Q such that $\mathcal{I}(Q) \cup \{\varnothing\}$ is not an $\bigcap$–structure on Q.

(iii) Show that, if Q is an ordered set, then $\langle \mathcal{I}(Q); \subseteq \rangle$ is a pre-CPO in which $\bigsqcup$ is given by set union, and $\alpha \colon x \mapsto {\downarrow}x$ is a (well-defined) order-embedding of Q into $\mathcal{I}(Q)$.

(iv) Prove that, if D is a domain, then the map $\varphi \colon a \mapsto D_a$, where $D_a := \{\, f \in F(D) \mid f \leqslant a \,\}$, is an order-isomorphism of D onto the set $\mathcal{I}(F(D))$ of ideals of $F(D)$.

3.20 Let Q be an ordered set and R a pre-CPO. Then R is called a **free pre-CPO generated by** Q if there is an order-preserving map $\eta \colon Q \to R$ such that for each order-preserving map $\varphi \colon Q \to P$, where P is a pre-CPO, there exists a unique continuous map $\varphi' \colon R \to P$ such that $\varphi' \circ \eta = \varphi$.

The pre-CPO $\langle \mathcal{I}(Q); \subseteq \rangle$ is called the **ideal completion** of Q and is denoted by $\mathbf{IC}(Q)$. By Exercise 3.19(iii), $\alpha \colon x \mapsto {\downarrow}x$ is an order-embedding of Q into the pre-CPO $\mathbf{IC}(Q)$.

(i) Show that, if both R and R' are free pre-CPOs generated by Q, then $R \cong R'$. [Hint. Use 1.13(4).]

(ii) By (i) we may, up to order-isomorphism, refer to *the* free pre-CPO R generated by Q. Show that $\eta \colon Q \hookrightarrow R$. [Hint. Try $\varphi = \alpha$ in the definition of R.]

(iii) Show that, if J is an ideal of Q, then $J = \bigsqcup_{x \in J} {\downarrow}x$ in $\mathbf{IC}(Q)$.

(iv) Show that $\mathbf{IC}(Q)$ is generated as a pre-CPO by $\alpha(Q)$, that is, the smallest sub-pre-CPO of $\mathbf{IC}(Q)$ containing $\alpha(Q)$ is $\mathbf{IC}(Q)$ itself.

(v) Prove that $\mathbf{IC}(Q)$ is the free pre-CPO generated by Q. [Hint. Take $\eta = \alpha$ and, given $\varphi \colon Q \to P$, define $\varphi' \colon \mathbf{IC}(Q) \to P$ by $\varphi' \colon J \mapsto \bigsqcup \varphi(J)$. Use Exercise 3.3 to show that φ' is continuous and (iii) above to show that φ' is unique.]

(vi) Show, directly and without reference to the ideal completion, that the uniqueness assumption on the continuous map φ' in the definition of the free pre-CPO generated by Q may be replaced by the assumption that R is generated as a pre-CPO by $\eta(Q)$.

3.21 Let Q be an ordered set and P a pre-CPO. The first part of the proof of Theorem 3.17 shows that $\langle Q \to P \rangle$ is a pre-CPO and the theorem itself implies that $[\mathbf{IC}(Q) \to P]$ is a pre-CPO. Show that $\varphi \mapsto \varphi'$ is an order-isomorphism from $\langle Q \to P \rangle$ onto $[\mathbf{IC}(Q) \to P]$.

3.22 Given a group G, describe an information system $\langle G, Con, \vdash \rangle$ whose set of elements is order-isomorphic to $\mathrm{Sub}\, G$ (cf. 3.34).

3.23 Let X be a set and let $F := \{\, Y \mid Y \Subset X \,\}$. Describe information systems $\mathbf{X}$ and $\mathbf{F}$, with X and F as their sets of tokens respectively, such that $|\mathbf{X}| = \wp(X)$ while $|\mathbf{F}| \cong \wp(X)$.

Define approximable mappings r from $\mathbf{X}$ to $\mathbf{F}$ and s from $\mathbf{F}$ to $\mathbf{X}$ such that the induced continuous maps $|r| : |\mathbf{X}| \to |\mathbf{F}|$ and $|s| : |\mathbf{F}| \to |\mathbf{X}|$, as defined in the proof of Proposition 3.49, are mutually inverse order-isomorphisms.

3.24   (i) Given an ordered set P, define an information system $\mathbf{P}$, with P as its token set, such that $|\mathbf{P}| = \mathcal{O}(P)$.

(ii) Let Q be an ordered set. Show that $\mathbf{Q} \trianglelefteq \mathbf{P}$ if and only if $Q \subseteq P$ and Q has the order induced from P.

(iii) Assume that $Q \subseteq P$ has the order induced from P. For $Y \Subset P$ and $q \in Q$ define

$$r : Y \rightsquigarrow b \iff b \leqslant a \text{ for some } a \in Y.$$

Show that r is an approximable mapping from $\mathbf{P}$ to $\mathbf{Q}$ and describe the corresponding continuous map from $\mathcal{O}(P)$ to $\mathcal{O}(Q)$.

3.25 Let $A = \mathbb{Q} \cap (0, 1)$. Define $\mathbf{A} = \langle A, Con, \vdash \rangle$ by setting $Y \in Con$ if and only if $Y \Subset A$ and $Y \vdash a$ if and only if $a \leqslant y$ for some $y \in Y$. Describe $F(|\mathbf{A}|)$ and show that the non-finite elements are in order-preserving correspondence with the half-open interval $(0, 1]$.

3.26 The triple $\mathbf{A} = \langle A, Con, \vdash \rangle$ is defined as follows:

(a) $A = \{\, (x_1, x_2) \times (y_1, y_2) \subseteq \mathbb{R} \times \mathbb{R} \mid x_1 < x_2 \,\&\, y_1 < y_2 \,\}$ (open rectangles in the plane),

(b) $Y \in Con$ if and only if $Y \Subset A \,\&\, \bigcap Y \neq \varnothing$,

(c) $Y \vdash a$ if and only if $\bigcap Y \subseteq a$.

Verify that $\mathbf{A}$ is an information system. Describe the finite elements and the maximal elements of $|\mathbf{A}|$.

3.27 Let $\mathbf{M} = \langle \mathbb{N} \times \mathbb{N}, Con, \vdash \rangle$, where

$Y \in Con \iff Y \in \mathbb{N} \times \mathbb{N}$ such that $(m_1, n_1), (m_2, n_2) \in Y$ and $m_1 \leqslant m_2$ imply $n_1 \leqslant n_2$,

$Y \vdash (m, n) \iff$

(a) $n = 1$ and $(\exists m_1 \in \mathbb{N}) \, m \leqslant m_1$ and $(m_1, 1) \in Y$, or

(b) $(\exists m_1, m_2 \in \mathbb{N}) \, m_1 \leqslant m \leqslant m_2$ and $(m_1, n), (m_2, n) \in Y$.

Show that $\mathbf{M}$ is an information system and that

$$|\mathbf{M}| = \{ \varphi \in (\mathbb{N} \longrightarrow \mathbb{N}) \mid \varphi \text{ is order-preserving} \}.$$

3.28 Let $\mathbf{F}$ be the information system given in Example 3.34(1) and let $\mathbf{M}$ be the information system from the previous example. Show that

$$\Sigma : Y \rightsquigarrow (m, n) \iff (1, k_1), \ldots, (m, k_m) \in Y \text{ and } \sum_{i=1}^{m} k_i = n$$

defines an approximable mapping from $\mathbf{F}$ to $\mathbf{M}$. Show that the corresponding continuous map σ from $|\mathbf{F}|$ to $|\mathbf{M}|$ satisfies $\sigma(f)(m) = \sum_{i=1}^{m} f(i)$ for each total map $f : \mathbb{N} \to \mathbb{N}$. What is the value of $\sigma(f)$ if f is a partial map and $1 \notin \operatorname{dom} \sigma$?

3.29 Let $(\mathfrak{L}, A)$ and $(\mathfrak{K}, B)$ be algebraic $\bigcap$-structures with $\mathfrak{L} \sqsubseteq \mathfrak{K}$. Define $\iota : \mathfrak{L} \to \mathfrak{K}$ by $\iota(S) := \bigcap \{ T \in \mathfrak{K} \mid S \subseteq T \}$ and $\pi : \mathfrak{K} \to \mathfrak{L}$ by $\pi(T) := T \cap A$ for all $S \in \mathfrak{L}$ and $T \in \mathfrak{K}$.

(i) Show that ι is well defined and that $\pi(\iota(S)) = S$ for all $S \in \mathfrak{L}$ and $\iota(\pi(T)) \subseteq T$ for all $T \in \mathfrak{K}$.

(ii) Show that π is continuous and ι is a continuous order-embedding.

3.30 Let $(\mathfrak{L}, A)$ be an $\bigcap$-structure and let $A^1 := (\{0\} \times A) \cup \{(1, 1)\}$. Construct an $\bigcap$-structure $\mathfrak{L} \boxplus \mathbf{1}$ on A^1 which is isomorphic as an ordered set to $\mathfrak{L} \oplus \mathbf{1}$.

Show that the domain constructor $\mathfrak{L} \mapsto \mathfrak{L} \boxplus \mathbf{1}$ is (a) $\sqsubseteq$-preserving, (b) continuous on base sets (and hence continuous).

Let $\mathfrak{L}^0 = \mathfrak{N}$, the unique $\bigcap$-structure based on $\varnothing$, and for all $n \in \mathbb{N}_0$ let $\mathfrak{L}^{n+1} := \mathfrak{L}^n \boxplus \mathbf{1}$. Construct $\mathfrak{L}^n$ for $n = 0, 1, 2, 3, 4$ and draw a labelled diagram of each. (For notational convenience let $a_0 = (1, 1)$ and $a_{n+1} = (0, a_n)$ for $n \geqslant 0$.) Indicate via arrows the embeddings $\iota_0 : \mathfrak{L}^0 \to \mathfrak{L}^1$, $\iota_1 : \mathfrak{L}^1 \to \mathfrak{L}^2$, etc. (see Exercise 3.29).

4
Fixpoint Theorems

Given an ordered set P and an order-preserving map $\Phi\colon P \to P$, does there exist a fixpoint for Φ? That is, does there exist a point $x \in P$ such that $\Phi(x) = x$? As stated, this is a purely order-theoretic problem, but much of the impetus to tackle it comes from further afield. The first section of this chapter presents the rudiments of fixpoint theory, both in the classic setting of complete lattices and, as most applications demand, in the setting of CPOs. Our introductory examples hint at this range of applications and reveal a connection with recursion. Our objective throughout is to give an overview of fixpoint theory and we do not attempt to discuss its specialized ramifications nor to do more than hint at the deeper computer science applications. The relationship between Zorn's Lemma and fixpoints is discussed in the second part of the chapter, which is aimed at those with some knowledge of set theory.

Fixpoint theorems and their applications

Much mathematical effort is expended on solving equations. These may be of very diverse types, but many important ones can be expressed in the form $\Phi(x) = x$, where $\Phi\colon X \to X$ is a map and any solution x is required to lie in X, which might be a set of real numbers, maps, or sets, or might be of some other type. A solution of such a **fixpoint equation**, when one exists, frequently has to be obtained by a process of successive approximation. Order theory has a role to play when X carries an order and when the required solution can be realized as the join of partial elements which approximate it. The domain equations we introduced in 3.52 can be cast as fixpoint equations. We now give other examples to indicate the range of circumstances in which fixpoint equations occur.

4.1 The factorial function. This example, a hoary chestnut in computer science texts, illustrates many features which recur elsewhere. The factorial function, **fact**, is the map on $\mathbb{N}_0$ given by $\mathbf{fact}(k) = k!$. It satisfies

$$\mathbf{fact}(k) = \begin{cases} 1 & \text{if } k = 0, \\ k\,\mathbf{fact}(k-1) & \text{if } k \geqslant 1. \end{cases}$$

To each map $f\colon \mathbb{N}_0 \to \mathbb{N}_0$ we may associate a new map $\overline{f}$ given by

$$\overline{f}(k) = \begin{cases} 1 & \text{if } k = 0, \\ kf(k-1) & \text{if } k \geqslant 1. \end{cases}$$

The equation satisfied by **fact** can then be recast in the form $\Phi(f) = f$, where $\Phi(f) = \overline{f}$.

To determine **fact**(k) for a given $k \geqslant 1$, we need to know **fact**$(k-1)$, and unless $k = 1$ this requires knowledge of **fact**$(k - 2)$, and so on. What we have here is a **recursive equation** satisfied by **fact** (recursive, from its latin roots, meaning running backwards). The entire factorial function cannot be unwound from its recursive specification in a finite number of steps. However we can, for each $n = 0, 1, \ldots$, determine in a finite number of steps the partial map f_n which is the restriction of **fact** to $\{0, 1, \ldots, n\}$; the graph of f_n is $\{(0, 1), (1, 1), \ldots, (n, n!)\}$. To accommodate approximations to **fact**, it is therefore natural to work not simply with maps from $\mathbb{N}_0$ to $\mathbb{N}_0$ but with all partial maps on $\mathbb{N}_0$. When this is done, we regard $\overline{f}$ as having $\{1\} \cup \{k \mid k - 1 \in \operatorname{dom} f\}$ as its domain. We may alternatively work with all maps from $\mathbb{N}_0$ to $(\mathbb{N}_0)_\perp$ (recall 1.29). When this is done, we take $\overline{f}(k) = \perp$ precisely when $f(k - 1) = \perp$. In summary, we may regard the factorial function as a solution of a recursive equation $\Phi(f) = f$, where $f \in (\mathbb{N}_0 \dashrightarrow \mathbb{N}_0)$ or equivalently $f \in (\mathbb{N}_0 \to (\mathbb{N}_0)_\perp)$.

In a similar way, we may consider the solution of $\Phi(f) = f$, where $\Phi(f) = \overline{f}$ as before, but now with $f \in (\mathbb{Z} \dashrightarrow \mathbb{Z})$. This equation does not have a unique solution. The values of f are uniquely determined on non-negative integers. However, $f(-1)$ may be assigned any value $\alpha \in \mathbb{Z}$. Each such assignment gives a solution of the fixpoint equation. Another solution is the partial map coinciding with **fact** on $\mathbb{N}_0$ and otherwise undefined.

4.2 Further examples: procedures as fixpoints.

(1) We may associate with every partial map π on $\mathbb{N}$ another such map $\Phi(\pi) = \overline{\pi}$, where

$$\overline{\pi}(k) = \begin{cases} 1 & \text{if } k = 1, \\ \pi(k - 1) + k & \text{if } k > 1. \end{cases}$$

Thus $\operatorname{dom} \overline{\pi} = \{1\} \cup \{k + 1 \mid k \in \operatorname{dom} \pi\}$. If π is a total map, so is $\overline{\pi}$. If π is undefined at i then $\overline{\pi}$ is undefined at $i + 1$. We seek possible solutions to the fixpoint equation $\Phi(\pi) = \pi$. Suppose π is a solution. Then certainly $\pi(1) = 1$ and, recursively, $\pi(k + 1) = \pi(k) + k + 1$. Hence π must be a total map and $\pi(k) = \sum_{i=1}^{k} i$, for all $k \in \mathbb{N}$, by an inductive proof. We can regard the solution to the equation $\Phi(\pi) = \pi$ as a program which, given k, outputs the sum of the first k natural numbers.

(2) **The while-loop.** In the examples above we started from a fixpoint equation and sought its solution(s). More commonly in applications, the starting point is a recursive definition of some map or procedure, with the object to be defined recognizable as a solution of some fixpoint equation. To illustrate, we return to the programming language fragment we used in 3.52 to introduce denotational semantics. We ask how commands of the form 'while B do C' should be assigned a meaning. Intuitively, the interpretation should be: 'so long as B is true, do C repeatedly; once B is false, stop in current state'. Thus in the notation of 3.52 we want, for any state σ, $\mathcal{C}[\![\texttt{while } B \texttt{ do } C]\!]\sigma$ to be $\mathcal{C}[\![\texttt{ while } B \texttt{ do } C]\!]\mathcal{C}[\![C]\!]\sigma$ if $\mathcal{B}[\![B]\!]$ is true and σ otherwise. More succinctly, we are demanding

$$\mathcal{C}[\![\texttt{while } B \texttt{ do } C]\!] = cond\,(\mathcal{B}[\![B]\!], \mathcal{C}[\![\texttt{while } B \texttt{ do } C]\!] \circ \mathcal{C}[\![C]\!], \mathrm{id}),$$

where $cond$ is as defined in 3.52 and $\mathrm{id}(\sigma) = \sigma$ for any state σ. This appears to require $\mathcal{C}[\![\texttt{while } B \texttt{ do } C]\!]$ to be defined in terms of itself. The position is clarified by writing A for $\mathcal{C}[\![\texttt{while } B \texttt{ do } C]\!]$ (a member of the domain $D := (St_\perp \to St_\perp)$) and Φ for the map from D to D which sends f to $cond\,(\mathcal{B}[\![B]\!], f \circ \mathcal{C}[\![C]\!], \mathrm{id})$. Then the while-loop equation becomes $\Phi(A) = A$. Thus the problem of showing that the recursive procedure 'while B do C' does have a proper definition is reduced to that of showing that a certain fixpoint equation has a solution.

We now begin to tackle the problem of the existence and construction of solutions to fixpoint equations, in an order-theoretic framework.

4.3 Technical note. Recall that we proved in 3.18 that for any $S \subseteq \mathbb{R}$ the set of maps from S to the flat CPO $S_\perp$ (or, equivalently, the set $(S \multimap S)$) is a CPO. Our preliminary examples therefore suggest that CPOs provide a suitable setting for our study. We need CPOs, not pre-CPOs, but (see the proof of 4.5) we do not require our maps to preserve $\perp$.

In the discussion of CPOs in Chapter 3 we needed to work with directed sets; here the emphasis is on chains. This allows us to by-pass directed sets and so reduce the dependence of this chapter on the last. We now take a CPO to be an ordered set with $\perp$ such that $\bigvee C$ (or, in the notation of 3.9, $\bigsqcup C$) exists in P for every non-empty chain C in P. The comments in 3.10 imply that this does not conflict with the definition in 3.9. Consequential amendments need to be made to other definitions, so that a continuous map becomes a map preserving joins of chains, etc.

4.4 Definitions. Let P be an ordered set and let $\Phi \colon P \to P$ be a map. We say $x \in P$ is a **fixpoint** of Φ if $\Phi(x) = x$. The set of all fixpoints of Φ is denoted by $\mathrm{fix}(\Phi)$; it carries the induced order. The least element of $\mathrm{fix}(\Phi)$, when it exists, is denoted by $\mu(\Phi)$.

The n-fold composite, Φ^n, of a map $\Phi \colon P \to P$ is defined as follows: Φ^n is the identity map if $n = 0$ and $\Phi^n = \Phi \circ \Phi^{n-1}$ for $n \geqslant 1$. If Φ is order-preserving, so is Φ^n.

Our first fixpoint theorem puts forward a candidate for the least fixpoint of an order-preserving map on a CPO, and confirms that the given construction always works when the map is continuous.

4.5 CPO Fixpoint Theorem I. *Let P be a CPO, let $\Phi \colon P \to P$ be an order-preserving map and define $\alpha := \bigsqcup_{n \geqslant 0} \Phi^n(\bot)$.*

(i) *If $\alpha \in \mathrm{fix}(\Phi)$, then $\alpha = \mu(\Phi)$.*

(ii) *If Φ is continuous then the least fixpoint $\mu(\Phi)$ exists and equals α.*

Proof. (i) Certainly $\bot \leqslant \Phi(\bot)$. Applying the order-preserving map Φ^n, we have $\Phi^n(\bot) \leqslant \Phi^{n+1}(\bot)$, for all n. Hence we have a chain

$$\bot \leqslant \Phi(\bot) \leqslant \ldots \leqslant \Phi^n(\bot) \leqslant \Phi^{n+1}(\bot) \leqslant \ldots$$

in P. Since P is a CPO, $\alpha := \bigsqcup_{n \geqslant 0} \Phi^n(\bot)$ exists. Let β be any fixpoint of Φ. By induction, $\Phi^n(\beta) = \beta$ for all n. We have $\bot \leqslant \beta$, whence we obtain $\Phi^n(\bot) \leqslant \Phi^n(\beta) = \beta$ by applying Φ^n. The definition of α forces $\alpha \leqslant \beta$. Hence if α is a fixpoint then it is the least fixpoint.

(ii) It will be enough to show that $\alpha \in \mathrm{fix}(\Phi)$. We have

$$\Phi\left(\bigsqcup_{n \geqslant 0} \Phi^n(\bot)\right) = \bigsqcup_{n \geqslant 0} \Phi(\Phi^n(\bot)) \quad \text{(since } \Phi \text{ is continuous)}$$

$$= \bigsqcup_{n \geqslant 1} \Phi^n(\bot)$$

$$= \bigsqcup_{n \geqslant 0} \Phi^n(\bot) \qquad \text{(since } \bot \leqslant \Phi^n(\bot) \text{ for all } n\text{).}\quad\blacksquare$$

4.6 Remarks. An instructive parallel may be drawn between Theorem 4.5 and another well-known fixpoint theorem. Banach's Contraction Mapping Theorem asserts that a contraction map on a complete metric space has a fixpoint. It serves to show the existence of solutions to a variety of equations, in particular differential equations. Further, a fixpoint can be explicitly constructed by an iterative process analogous to that in 4.5. The approximating sequence converges because the metric space is complete, and its limit is a fixpoint thanks to the continuity of the map. In 4.5 we simply have order-theoretic notions of completeness and continuity replacing the topological ones in Banach's Theorem.

In many practical applications of 4.5, the map Φ is defined on a CPO of the form $(S \multimap S)$, where $S \subseteq \mathbb{R}$. We noted in 3.18 that $(S \multimap S)$ is isomorphic to $(S \to S_\perp)$ and also to the strict maps in $[S_\perp \to S_\perp]$. The bottom element of $(S \multimap S)$ is $\varnothing$; this corresponds to the map in $(S \to S_\perp)$ or in $[S_\perp \to S_\perp]$ which sends every element to $\perp$ in $S_\perp$. We denote this map by $\perp$ in either case.

Let $P = (S \multimap S)$. When Φ is to be defined on P, we shall always take $\Phi(f)$ to have the maximum possible domain set. In other words, we take $\operatorname{dom} \Phi(f)$ to consist of all points x for which $(\Phi(f))(x)$ makes sense. The map $\Phi \colon P \to P$ is order-preserving if and only if graph $f \subseteq$ graph g implies graph $\Phi(f) \subseteq$ graph $\Phi(g)$ for any $f, g \in P$. When this condition is satisfied, $\{\Phi^n(\perp)\}_{n \geqslant 0}$ forms a consistent set of partial maps with successively bigger domain sets. Further, graph $(\bigsqcup_{n \geqslant 0} \Phi^n(\perp)) = \bigcup_{n \geqslant 0} \operatorname{graph} \Phi^n(\perp)$.

The advantage of identifying $(S \multimap S)$ with the strict maps in $[S_\perp \to S_\perp]$ is that members of the latter set can be combined by functional composition. With this viewpoint it is often possible to recognize $\Phi(f)$ as a composite of f with order-preserving maps on flat CPOs. It is then immediate that Φ is order-preserving.

4.7 Examples revisited. The recursive specifications for the factorial function, **fact**, and for the while-loop come within the scope of Theorem 4.5. In 4.1, we recognized **fact** as the solution of a fixpoint equation $\Phi(f) = f$ for $f \in (\mathbb{N}_0 \to (\mathbb{N}_0)_\perp)$, where

$$(\Phi(f))(k) = \begin{cases} 1 & \text{if } k = 0, \\ k f(k-1) & \text{if } k \geqslant 1. \end{cases}$$

It is obvious that Φ is order-preserving. We have graph $\Phi(\perp) = \{(0,1)\}$, graph $\Phi^2(\perp) = \{(0,1),(1,1)\}$, etc. An easy induction confirms that $f_n = \Phi^n(\perp)$ for all n, where $\{f_n\}$ is the sequence of partial maps defined in 4.1 as approximations to **fact**. Forming the join, by taking the union of the graphs, we see that $\bigsqcup_{n \geqslant 0} \Phi^n(\perp)$ is the map **fact**: $k \mapsto k!$ on $\mathbb{N}_0$. This is the least fixpoint of Φ, whether we work in $(\mathbb{N}_0 \multimap \mathbb{N}_0)$ or in $(\mathbb{Z} \multimap \mathbb{Z})$. Over $\mathbb{Z}$, there are other solutions to $\Phi(f) = f$ as well, but these contain extraneous unforced information. This illustrates that it is the *least* fixpoint which is likely to provide the computationally most natural solution to a recursive equation for a map or procedure.

The while-loop example, 4.2(2), shows striking similarities to the factorial one. The successive approximations to $\mathcal{C}[\![\texttt{while } B \texttt{ do } C]\!]$ are given, for $n \geqslant 1$, by $\{W_n(B,C)\}_{n \geqslant 1}$, where $W_n(B,C)$ coincides with 'while B do C' for computations involving fewer than n iterations of the loop, and is undefined otherwise; $W_0(B, C)$ is 'do nothing'. The transition from $W_n(B, C)$ to $W_{n+1}(B, C)$ accomplishes one stage in unwinding the loop. As in the previous example, the fixpoint theorem allows us to realize a recursively defined object as the limit of partially defined objects which can be specified without recursion.

4.8 Further examples.

(1) Consider the order-preserving map $\Phi \colon (\mathbb{N} \to \mathbb{N}_\perp) \to (\mathbb{N} \to \mathbb{N}_\perp)$ given by

$$(\Phi(f))(k) = \begin{cases} 1 & \text{if } k = 1, \\ 2f(k-1) - 1 & \text{otherwise.} \end{cases}$$

We have $\operatorname{graph} \Phi(\perp) = \{(1, 1)\}$ and $\operatorname{graph} \Phi^2(\perp) = \{(1, 1), (2, 1)\}$ and so on. Induction establishes that $\Phi^n(\perp)$ is the partial map taking the constant value 1 on $\{1, 2, \ldots, n\}$ (and undefined otherwise). We conclude that the least fixpoint of Φ has graph $\mathbb{N} \times \{1\}$. Hence the equation $\Phi(f) = f$ has a unique solution, given by $f(k) = 1$ for all $k \in \mathbb{N}$. This shows that very simple maps may hide behind complicated recursive specifications.

(2) The fixpoint formula in Theorem 4.5 may be used to determine certain recursively defined maps of more than one variable. Consider the following equation on $(\mathbb{N}^2 \to (\mathbb{N})_\perp)$:

$$f(j, k) = \begin{cases} 1 & \text{if } j = k, \\ (k+1)f(j, k+1) & \text{otherwise.} \end{cases}$$

We take $\Phi \colon (\mathbb{N}^2 \to \mathbb{N}_\perp) \to (\mathbb{N}^2 \to \mathbb{N}_\perp)$ to be the map taking f to the right-hand side of the equation above; Φ is order-preserving. The domain of $\Phi(f)$ is $\{(j, k) \mid j = k \text{ or } (j, k+1) \in \operatorname{dom} f\}$. A rather complicated induction on n shows that the approximations to the least fixpoint of Φ are given by

$$\Phi^n(\perp)(j, k) = \begin{cases} j!/k! & \text{if } 0 \leqslant j - k \leqslant n, \\ \perp & \text{otherwise.} \end{cases}$$

We deduce that the least fixpoint is undefined at (j, k) if $k > j$ and takes value $j!/k!$ there if $k \leqslant j$.

(3) Let Σ^{**} be the set of all binary strings. Given a finite string, u, and an arbitrary string, v, we denote by uv the string obtained by concatenating u and v. Let $\Phi(u) = 01u$ for $u \in \Sigma^{**}$. It is intuitively clear that the fixpoint equation $\Phi(u) = u$ has a unique

solution, namely α, the infinite string of alternating zeros and ones. This is exactly the solution we obtain by taking the empty string, $\varnothing$, and forming $\bigsqcup_{n \geqslant 1} \Phi^n(\varnothing)$. Clearly, $\Phi^n(\perp)$ is the $2n$-element string $0101\ldots01$. The join in Σ^{**} of these strings is α, which is certainly a fixpoint of Φ. Trivially Φ is order-preserving so, by Theorem 4.5(i), α is indeed the least fixpoint.

4.9 Recursively defined domains. Our examples so far in this chapter have concerned maps and procedures. We now hint at the potential of CPO Fixpoint Theorem I for creating solutions to domain equations of the kind we introduced in 3.52. As a simple example, consider the domain equation $\mathbf{F}(P) \cong P$, where $\mathbf{F}(P) = P_\perp$. Starting from $\mathbf{1}$ and forming $\{\mathbf{F}^n(\mathbf{1})\}_{n \geqslant 1}$, we obtain $\mathbf{1}_\perp$, $\mathbf{1}_\perp \cong \mathbf{2}$, $\mathbf{2}_\perp \cong \mathbf{3}$, etc. This makes it highly plausible that the least solution to $\mathbf{F}(P) \cong P$ is the domain $\mathbb{N} \oplus \mathbf{1}$. To confirm this, we must realize $\mathbf{F}$ as an order-preserving map on a CPO of domains and verify that $\mathbb{N} \oplus \mathbf{1}$ is indeed the join in this CPO of the approximations $\{\mathbf{F}^n(\mathbf{1})\}_{n \geqslant 0}$. We do not have an ordering of 'abstract' domains and must therefore realize domains concretely, either as $\bigcap$–structures (ordered by $\sqsubseteq$) or as information systems (ordered by $\trianglelefteq$). Using $\bigcap$–structures, we then have to construct $\mathbf{F}^n(\mathfrak{N})$. The reward for disentangling this notational horror (recall 3.46) is that the correct embeddings are then given at once via Exercise 3.29. This confirms that the 'natural' nesting of the chains $\mathbf{1}, \mathbf{2}, \ldots$ (in which each chain sits, in $\mathbb{N} \oplus \mathbf{1}$, as a down-set in the next) corresponds to the $\sqsubseteq$–ordering of their realizations as $\bigcap$–structures. See Figure 4.1 (in which $\mathbf{0}$ is used as an abbreviation for $(0,0)$).

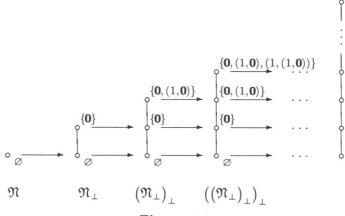

Figure 4.1

As a second example, consider the equation $P \cong P \oplus_{\perp} P$. Let $\mathbf{F}(P) = P \oplus_{\perp} P$. Then the least fixpoint of $\mathbf{F}$ is isomorphic to Σ^{**}, the domain of all binary strings. The nth approximation, $\mathbf{F}^n(\varnothing)$, to the solution is the set of strings of length at most n; again the ordering $\sqsubseteq$ is the 'natural' nesting. Finally we note that we cannot exhibit explicit solutions to domain equations, such as $P \cong (P \to P)_{\perp}$, which involve function spaces. As those who did Exercise 1.17 will realize, these spaces get very unwieldy very quickly and the limit domain cannot be visualized easily. However, so long as the domain constructor used is continuous, we know that there is a solution to the associated fixpoint equation, and this is what really matters, since it ensures that the models required for denotational semantics do indeed exist. Invoking CPO Fixpoint Theorem III below, we can even replace 'continuous' by 'order-preserving' above.

4.10 Remarks: the role of continuity. Consider the chain $P = \mathbb{N} \oplus \mathbf{2}$ and define Φ to be the (order-preserving) map fixing $\top$ and taking every other element to its upper cover. This map has $\top$ as its unique fixpoint, but $\top \neq \bigsqcup_{n \geqslant 0} \Phi^n(\perp)$. This example shows that for a non-continuous map Φ on a CPO we cannot expect $\bigsqcup \Phi^n(\perp)$ necessarily to provide a fixpoint.

Only for a map which is continuous does Theorem 4.5 guarantee the existence of a fixpoint. In simple cases it is possible to side-step the question of continuity by appealing to the first part of the theorem. This is convenient since verifying continuity can be a non-trivial undertaking, especially when the underlying CPO is a set of maps or the map Φ a domain constructor. It is much easier to decide whether a map preserves order, and we are led to ask whether an order-preserving map on a CPO must have a fixpoint. This turns out to be true though we have to work hard to establish it.

Before embarking on this proof we derive a classic fixpoint theorem due to Knaster and Tarski. This theorem is important in its own right and for the clues it provides to a systematic search for fixpoints more generally.

4.11 The Knaster–Tarski Fixpoint Theorem. *Let L be a complete lattice and $\Phi \colon L \to L$ an order-preserving map. Then*

$$\bigvee \{\, x \in L \mid x \leqslant \Phi(x) \,\} \in \mathrm{fix}(\Phi).$$

Proof. Let $H = \{\, x \in L \mid x \leqslant \Phi(x) \,\}$ and $\alpha = \bigvee H$. For all $x \in H$ we have $x \leqslant \alpha$, so $x \leqslant \Phi(x) \leqslant \Phi(\alpha)$. Thus $\Phi(\alpha) \in H^u$, whence $\alpha \leqslant \Phi(\alpha)$.

We now use this inequality to prove the reverse one (!) and thereby complete the proof that α is a fixpoint. Since Φ is order-preserving, $\Phi(\alpha) \leqslant \Phi(\Phi(\alpha))$. This says $\Phi(\alpha) \in H$, so $\Phi(\alpha) \leqslant \alpha$. ∎

4.12 Remark. The Knaster–Tarski Theorem shows that any order-preserving map on a powerset has a fixpoint. One such application yields as a by-product the famous Schröder–Bernstein Theorem stating that there is a bijection between sets A and B if there exist one-to-one maps from A to B and from B to A. For the proof, see Exercise 4.12. A full discussion of the Schröder-Bernstein Theorem and its uses in set theory can be found in [8].

The formula in Theorem 4.11 constructs what is easily seen to be the *greatest* fixpoint of Φ. A dual version of the proof produces $\mu(\Phi)$. Notice that both of the fixpoint theorems 4.5 and 4.11 find a fixpoint in the set $\{\, x \mid x \leqslant \Phi(x) \,\}$ of **pre-fixpoints**. The lemma below extracts the ideas underpinning 4.11.

4.13 Lemma. *Let P be an ordered set and let $\Phi\colon P \to P$ be a map. Let $H = \{\, x \in P \mid x \leqslant \Phi(x) \,\}$.*

(i) *Suppose $A \subseteq H$ and $\Phi(A) \subseteq A$. Then any maximal element of A belongs to* $\mathrm{fix}(\Phi)$.

(ii) *If Φ is order-preserving, then $\Phi(H) \subseteq H$ and*

    (a) $\bigvee_P H \in \mathrm{fix}(\Phi)$ *if this exists,*

    (b) *any maximal element of H belongs to* $\mathrm{fix}(\Phi)$.

Proof. (i) Assume α is a maximal point of A. Since $A \subseteq H$, we have $\alpha \leqslant \Phi(\alpha)$. We deduce that $\Phi(\alpha) = \alpha$, because $\Phi(\alpha) \in A$ and α is maximal in A.

(ii) The first two assertions are proved by the same techniques as were used in the proof of 4.11. For the last part take $A = H$ in (i). ∎

The next theorem is by far the hardest in this chapter, but yields substantial dividends later. It will enable us to prove easily that any order-preserving map on a CPO has a fixpoint and provides the key to our treatment of Zorn's Lemma in the next section.

4.14 CPO Fixpoint Theorem II. *Let P be a CPO and let $\Phi\colon P \to P$ be a map such that $x \leqslant \Phi(x)$ for all $x \in P$. Then Φ has a fixpoint.*

Proof. Lemma 4.13(ii) suggests a strategy for the proof. We seek a subset C of P such that $\Phi(C) \subseteq C$ (we shall call any such set Φ-**invariant**) and such that C has a maximal point. We know $\bigsqcup C$ exists in P for any non-empty chain. We therefore seek to choose C to be a non-empty chain which is Φ-invariant and which is a sub-CPO of P. Then the supremum of C actually lies *in* C (and thus is a maximal point of C). To build such a chain we might try to start from $\perp$ and add (towards achieving Φ-invariance) successively $\Phi(\perp)$, $\Phi(\Phi(\perp))$,.... If this countable chain has a maximal element, we are home. Failing that, the chain does at least have a supremum γ and we may start climbing afresh, adding γ, $\Phi(\gamma)$, $\Phi(\Phi(\gamma))$, ... The problem is to ensure that this process ever terminates. The formal proof starts from the other end, by taking a minimal Φ-invariant sub-CPO of P and showing that it is in fact a chain of the kind envisaged above.

Accordingly we let

$$\mathcal{E} := \{\, A \subseteq P \mid \Phi(A) \subseteq A \text{ and } A \text{ is a sub-CPO of } P\}$$

and define $C = \bigcap\{\, A \mid A \in \mathcal{E}\,\}$. Then C is a sub-CPO of P, by 3.12, and is easily seen to be Φ-invariant. If $A \in \mathcal{E}$ and $A \subseteq C$, then $A = C$. This observation is applied twice. On both occasions we take A to be those elements of C which satisfy some property we wish to establish for all of C. We then show that this set is in $\mathcal{E}$. (This is exactly the idea behind proofs by induction: a subset of $\mathbb{N}$ containing 1 and invariant under the successor function is the whole of $\mathbb{N}$.)

We wish to show C is a chain. This means

$$(\forall x \in C)((\forall y \in C)\, y \in {\downarrow}x \cup {\uparrow}x).$$

Because $x \leqslant \Phi(x)$ for all $x \in P$, this is certainly satisfied if C equals

$$A_x := \{\, y \in C \mid y \in {\downarrow}x \cup {\uparrow}\Phi(x)\,\}$$

for every $x \in C$. The strategy now is to define a set $A \subseteq C$ such $A \in \mathcal{E}$ and such that $x \in A$ implies that $A_x \in \mathcal{E}$. (Then we will be able to deduce that $A = C$ and thence that $A_x = C$ for all $x \in C$, as required.)

To see how to define A, we look at what is needed to ensure $A_x \in \mathcal{E}$. It is easy to check that A_x is a sub-CPO of P. What about Φ-invariance? Fix $x \in C$. To show that $y \in A_x$ implies $\Phi(y) \in A_x$ we require

(i) $y < x$ implies $\Phi(y) \leqslant x$ or $\Phi(y) \geqslant \Phi(x)$,

(ii) $y = x$ implies $\Phi(y) \leqslant x$ or $\Phi(y) \geqslant \Phi(x)$, and

(iii) $y \geqslant \Phi(x)$ implies $\Phi(y) \leqslant x$ or $\Phi(y) \geqslant \Phi(x)$.

Here (ii) is automatic and so is (iii), since $\Phi(y) \geqslant y$ for any $y \in P$; but (i) is not. We define A so that (i) is true if $x \in A$:

$$A = \{\, x \in C \mid (\forall y \in C)\,(y < x \Rightarrow \Phi(y) \leqslant x)\,\}.$$

Summing up, we have $A_x \in \mathcal{E}$ whenever $x \in A$, and it remains to show that $A \in \mathcal{E}$, to obtain $A = C$. Our intuitive picture of C as a chain in which each point is covered by its Φ-image encourages us to hope that this is indeed true.

First consider Φ-invariance. Fix $x \in A$. By construction, $A_x = C$. Let $y \in C$ and $y < \Phi(x)$. We want $\Phi(y) \leqslant \Phi(x)$. We know $y \in A_x$, so either $y \leqslant x$ or $y \geqslant \Phi(x)$. If $y < x$, then $\Phi(y) \leqslant x$, because $x \in A$. In this case $\Phi(y) \leqslant x \leqslant \Phi(x)$, so $\Phi(y) \leqslant \Phi(x)$. If $y = x$, we have $\Phi(y) \leqslant \Phi(x)$, trivially. The final case, $y \geqslant \Phi(x)$, doesn't arise as it leads to $\Phi(y) \geqslant y \geqslant \Phi(x) > y$, $\frac{1}{2}$.

We now show that A is a sub-CPO of P. Vacuously, $\bot \in A$. Let K be a non-empty chain in A. To show $\bigsqcup K \in A$ we take $z \in C$ with $z < \bigsqcup K$ and check that $\Phi(z) \leqslant \bigsqcup K$. We know $A_k = C$ for all $k \in K$, so that either

(a) $\Phi(k) \leqslant z$ for all $k \in K$, or

(b) there exists $k_1 \in K$ with $z \leqslant k_1$.

In case (a), $k \leqslant \Phi(k) \leqslant z < \bigsqcup K$ for all $k \in K$, $\frac{1}{2}$. In case (b), $z < k_1$ implies $\Phi(z) \leqslant k_1 \leqslant \bigsqcup K$ (since $k_1 \in A$), and $z = k_1$ implies $k_1 < \bigsqcup K$ whence $\Phi(z) = \Phi(k_1) \leqslant \bigsqcup K$ (since $A_{k_1} = C$).

The highly convoluted argument above shows that C is indeed a chain. Let $\alpha = \bigvee C$. Since C is a sub-CPO of P, we have $\alpha \in C$. Lemma 4.13(i) implies that α is indeed a fixpoint of Φ. ∎

4.15 CPO Fixpoint Theorem III. *Let P be a CPO and $\Phi: P \to P$ an order-preserving map. Then Φ has a least fixpoint.*

Proof. If A is a Φ-invariant sub-CPO of P on which $x \leqslant \Phi(x)$ for all x, then Theorem 4.14 applies to A and Φ restricted to A to show the existence of a fixpoint of Φ. Let

$$A = \{\, x \in P \mid x \leqslant \Phi(x) \text{ and } x \in \mathrm{fix}(\Phi)^\ell \,\}.$$

(If $\mathrm{fix}(\Phi)$ were in fact empty, the second condition would be vacuous.) The set A certainly contains $\bot$. Now let C be a chain in A and denote $\bigsqcup_P C$ by z. Then we have $x \leqslant \Phi(x) \leqslant \Phi(z)$, for all $x \in C$. This makes $\Phi(z)$ an upper bound for C, so that $z \leqslant \Phi(z)$. Also, for any $\gamma \in \mathrm{fix}(\Phi)$ we have $x \leqslant \gamma$ for all $x \in C$. Hence $z \in A$ and we conclude that A is a sub-CPO of P. It only remains to show Φ-invariance. This follows from Lemma 4.13(ii) and the observation that, for any $x \in A$ and any $\beta \in \mathrm{fix}(\Phi)$, we have $x \leqslant \beta$ and so $\Phi(x) \leqslant \Phi(\beta) = \beta$.

Let γ be a fixpoint of Φ restricted to A. Then $\gamma \in \mathrm{fix}(\Phi)^\ell$, and hence $\gamma = \mu(\Phi)$. ∎

4.16 Stocktaking. We now have three theorems guaranteeing the existence of a least fixpoint:

CPO Fixpoint Theorem I (4.5) for a continuous map on a CPO,

the Knaster–Tarski Theorem (4.11) for an order-preserving map on a complete lattice, and

CPO Fixpoint Theorem III (4.15) for an order-preserving map on a CPO.

The first two (which have stronger hypotheses than the last) both provide a formula for the least fixpoint. Those readers who know about ordinals may have surmised from the discussion introducing the proof of Theorem 4.14 that it is possible to extend the formula in 4.5 by taking a limit over a chain indexed by ordinals. This is indeed the case; an outline can be found in Exercise 4.13.

Theorems 4.5 and 4.11 are in a sense optimal. The following converses exist. Both are difficult to prove; see [38].

4.17 Theorem. *Let P be an ordered set.*

(i) *If P is a lattice and every order-preserving map $\Phi \colon P \to P$ has a fixpoint, then P is a complete lattice.*

(ii) *If every order-preserving map $\Phi \colon P \to P$ has a least fixpoint, then P is a CPO.*

We have already remarked that where fixpoint theory is used to solve recursive equations, the least fixpoint (if it exists) is likely to provide the required solution. Proposition 4.18 gives additional information about the location of $\mu(\Phi)$ and about the set of all fixpoints of an order-preserving map on a CPO. For the proof see Exercises 4.14 and 4.15.

4.18 Proposition. *Let P be a CPO and assume that $\Phi \colon P \to P$ is an order-preserving map. Then*

(i) $\mu(\Phi) \in \{\, x \in P \mid x \leqslant \Phi(x) \,\}^u$;

(ii) $\mathrm{fix}(\Phi)$ *is a CPO;*

(iii) *if P is a complete lattice, so is $\mathrm{fix}(\Phi)$.*

4.19 Reasoning with fixpoints. We give a final group of examples to illustrate how information can be gleaned about recursively defined objects. One of computer science's thornier topics is program verification: proving mathematically that given programs terminate and that they output what is demanded of them. For procedures specified as least fix-points several techniques are available for handling such problems. Our examples are designed only to hint at the possibilities. Readers seeking a full working knowledge of the techniques are referred to the specialist texts listed in the Appendix.

We have seen in 4.7 and 4.8 that induction can be used to establish what a least fixpoint outputs for a given input. There is a different version of induction, tailor-made for fixpoint theory, which is often more convenient to apply than mathematical induction. Let Φ be a continuous map on a CPO P and let $S \subseteq P$ satisfy:

(FI)$_1$ $\perp \in S$;

(FI)$_2$ $x \in S$ implies $\Phi(x) \in S$;

(FI)$_3$ for any chain $x_0 \leqslant x_1 \leqslant x_n \leqslant \ldots$ in S, we have $\bigsqcup_{n \geqslant 0} x_n \in S$.

Then, by CPO Fixpoint Theorem I, $\mu(\Phi) \in S$. We refer to this as the **Principle of Fixpoint Induction**.

To illustrate, let Φ be a continuous map on a CPO P and let Ψ and Θ be continuous maps on P such that $\Phi \circ \Psi = \Psi \circ \Phi$, $\Phi \circ \Theta = \Theta \circ \Phi$ and $\Psi(\perp) = \Theta(\perp)$. We can prove by mathematical induction that $\Psi(\Phi^n(\perp)) = \Theta(\Phi^n(\perp))$ for all n. Since Ψ and Θ are continuous, and $\mu(\Phi) = \bigsqcup \Phi^n(\perp)$ by Theorem 4.5, we deduce that $\Psi(\mu(\Phi)) = \Theta(\mu(\Phi))$. Alternatively we can argue as follows. Define $S := \{\, x \in P \mid \Psi(x) = \Theta(x)\,\}$. Then (FI)$_1$ holds by hypothesis, (FI)$_2$ because Φ commutes with each of Ψ and Θ, and (FI)$_3$ because Ψ and Θ are continuous.

In general the set S in the Principle of Fixpoint Induction is taken to be a set on which some property desired of the least fixpoint holds. It will be of the form $S = \{\, x \in P \mid \mathbf{P}(x)\,\}$, where $\mathbf{P}$ is some predicate (that is, some statement taking value true or false). When P is a CPO of partial maps, we might, for example, have $\mathbf{P}(f)$ as the assertion that f is not defined at some given point c (or, alternatively, that a computation to determine $f(c)$ does not terminate). As an example, consider $P = (\mathbb{N}^2 \to \mathbb{N}_\perp)$ and

$$\overline{f}(j,k) = (\Phi(f))(j,k) := \begin{cases} 1 & \text{if } j = k, \\ (k+1)(k+2)f(j,k+2) & \text{otherwise.} \end{cases}$$

We claim that the associated least fixpoint, $\mu(\Phi)$, is undefined at (j,k) if $j < k$. To check this, take

$$S = \{\, g \in P \mid j < k \implies g(j,k) = \perp \,\}$$

and show that (FI)$_1$, (FI)$_2$ and (FI)$_3$ are satisfied. This technique of fixpoint induction can be adapted to a wide range of circumstances, but does have its limitations, since not all predicates yield a set S on which the Principle of Fixpoint Induction works.

In a different direction, we note that Proposition 4.18(i) potentially gives information about $\mu(\Phi)$. Assume we can exhibit some $x \in P$ such that $x \leqslant \Phi(x)$. Then we must have $x \leqslant \mu(\Phi)$.

Finally we mention a technique which is extremely useful for showing the equivalence of program constructs in denotational semantics, for example different versions of 'while'. Let Φ and Ψ be maps on an ordered set P which possess least fixpoints. If $\mu(\Phi)$ is a fixpoint of Ψ and $\mu(\Psi)$ is a fixpoint of Φ, we have $\mu(\Psi) \leqslant \mu(\Phi)$ and $\mu(\Phi) \leqslant \mu(\Psi)$ so that $\mu(\Phi) = \mu(\Psi)$. This provides a method for showing the equality of two objects defined by least fixpoints: show that each is a fixpoint of the map associated with the other.

The existence of maximal elements and Zorn's Lemma

There are many examples in mathematics of statements which, overtly or covertly, assert the existence of an element maximal in some ordered set (commonly, a family of sets under inclusion). Such results belong naturally in a course on the foundations of mathematics since they rely on Zorn's Lemma and this has its roots in the axiomatic basis for set theory. It would be inappropriate to include here a full discussion of the role and status in mathematics of the many equivalents of Zorn's Lemma, among which is the Axiom of Choice. We seek to complement the treatment in set theory texts of this important topic and, although this section is self-contained, it is principally directed at those who have previously encountered the Axiom of Choice.

Aside from the treatment of ordinals, ordered sets have traditionally played a peripheral role in introductory set theory courses. It is not unusual for ordered sets only to be formally introduced immediately before Zorn's Lemma is presented. Our aim is to set Zorn's Lemma in its order-theoretic context. En route, we are able to prove some interesting results about ordered sets themselves, and to provide the set-theoretic basis for parts of Chapters 9 and 10. We also hope to refute the view prevalent amongst some mathematicians that Zorn's Lemma is psychologically unappealing.

4.20 The Axiom of Choice. A non-empty ordered set may, but need not, possess maximal elements (see 1.22 for examples). Consider the following argument, purporting to show that a given CPO P has a maximal element. Suppose every element of P fails to be maximal.

This means that, for each $x \in P$, the set $\{\, y \in P \mid y > x \,\}$ is non-empty. For each x select a point $y > x$. Since y depends on x, we label it $\Phi(x)$. Then $x \mapsto \Phi(x)$ defines a map Φ on P, to which we may apply CPO Fixpoint Theorem II (4.14). Since $x < \Phi(x)$ for every $x \in P$, we have a contradiction, so P does indeed have a maximal element. We should come out in the open and make it clear that in the above argument we invoke the Axiom of Choice. This asserts that it is possible to find a map which picks one element from each member of a family of sets (only for finitely many sets can this be done without recourse to an axiom additional to those used in standard ZF set theory). Formally, the Axiom of Choice may be stated as follows:

(AC) Given a non-empty family $\mathcal{A} = \{A_i\}_{i \in I}$ of non-empty sets, there exists a **choice function for $\mathcal{A}$**, that is, a map

$$ f \colon I \to \bigcup_{i \in I} A_i \text{ such that } (\forall i \in I)\, f(i) \in A_i. $$

To apply (AC) to obtain the map required above we take $I = P$ and $A_x = \{\, y \in P \mid y > x \,\}$ for each $x \in P$.

The same sleight of hand occurred in the derivation of Theorem 2.26. This result also relies on (AC), which may be applied to produce the required chain.

4.21 Maximality axioms.   In the same way that (AC) may be regarded as an (optional) axiom of set theory, and deductions made from it, we may take the following statement as a postulate.

(ZL)   Let P be a non-empty ordered set in which every chain has an upper bound. Then P has a maximal element.

We shall also need three other axioms concerning the existence of a maximal element. These are as follows.

(ZL)$'$ Let $\mathcal{E}$ be a non-empty family of sets such that $\bigcup_{i \in I} A_i \in \mathcal{E}$ whenever $\{A_i\}_{i \in I}$ is a non-empty chain in $\langle \mathcal{E}; \subseteq \rangle$. Then $\mathcal{E}$ has a maximal element.

(ZL)$''$ Let P be a CPO. Then P has a maximal element.

(KL)   Let P be an ordered set. Then every chain in P is contained in a maximal chain.

Clearly (ZL)$'$ is just the restriction of (ZL) to families of sets. Our next theorem shows that the five assertions (AC), (ZL), (ZL)$'$, (ZL)$''$

and (KL) are all equivalent. It is the implication (AC) $\Rightarrow$ (ZL) that we refer to as **Zorn's Lemma**. (Some authors use Zorn's Lemma to mean the statement (ZL) instead.) Similarly, the implication (AC) $\Rightarrow$ (KL) is **Kuratowski's Lemma**.

4.22 Theorem. *Each of the statements below implies the other four.*

(AC) *Every non-empty family of non-empty sets has a choice function.*

(ZL) *Every non-empty ordered set in which every chain has a upper bound possesses a maximal element.*

(ZL)′ *Every non-empty family $\mathcal{E}$ of sets such that $\bigcup \mathcal{C} \in \mathcal{E}$ for every non-empty chain $\mathcal{C}$ in $\mathcal{E}$ possesses a maximal element.*

(ZL)″ *Every CPO has a maximal element.*

(KL) *In an ordered set, every chain is contained in a maximal chain.*

Proof. We prove (AC) $\Rightarrow$ (ZL)″ $\Rightarrow$ (KL) $\Rightarrow$ (ZL) $\Rightarrow$ (ZL)′ $\Rightarrow$ (AC). The first implication has been obtained in 4.20 and the fourth is trivial since (ZL)′ is a restricted form of (ZL).

We next prove that (ZL)″ implies (KL). Take an ordered set P and let $\mathcal{P}$ denote the family of all chains in P, including the empty chain, and order this family of sets by inclusion. We claim that $\mathcal{P}$ is a CPO. It has the empty chain as its bottom element. Now let $\mathcal{C} = \{C_i\}_{i \in I}$ be a chain in $\mathcal{P}$. Let $C = \bigcup_{i \in I} C_i$. We claim that C is a chain. Let $x, y \in C$. We require to show that x and y are comparable. There exist $i, j \in I$ such that $x \in C_i$ and $y \in C_j$. Since $\mathcal{C}$ is a chain, we have $C_i \subseteq C_j$ or $C_j \subseteq C_i$. Assume, without loss of generality, that $C_i \subseteq C_j$. Then x, y both belong to the chain C_j, and hence x and y are comparable. We may therefore apply (ZL)″ to $\mathcal{P}$ to obtain a maximal element C^* in $\mathcal{P}$.

The next step is to show that (KL) implies (ZL). Let P be a non-empty ordered set in which every chain has an upper bound. By (KL), an arbitrarily chosen chain in P is contained in a maximal chain, C^*, say. By hypothesis, C^* has an upper bound u in P. If u were not a maximal element of P, we could find $v > u$. Clearly $v \notin C^*$, since $u \geqslant c$ for all $c \in C^*$. Thus $C^* \cup \{v\}$ would be a chain strictly containing the maximal chain C^*, ⨍.

Finally, we prove (ZL)′ implies (AC). Consider the ordered set P of partial maps from I to $\bigcup_{i \in I} A_i$ (cf. 1.6). By identifying maps with their graphs we may regard P as a family of sets ordered by inclusion. Let $\mathcal{E} = \{\pi \in P \mid (\forall i \in \text{dom}\,\pi)\,\pi(i) \in A_i\}$. Certainly $\mathcal{E} \neq \varnothing$ since the partial map with empty domain vacuously belongs to $\mathcal{E}$. Now let

$\mathcal{C} = \{\pi_j\}_{j \in J}$ be a chain in $\mathcal{E}$. Because $\mathcal{C}$ is a chain, the partial maps π_j are consistent and the union of their graphs is the graph of a partial map, which necessarily belongs to $\mathcal{E}$. By (ZL)', $\mathcal{E}$ has a maximal element, $f: \operatorname{dom} f \to \bigcup A_i$, say. Provided f is a total map, it serves as the required choice function. Suppose f is not total, so that there exists $k \in I \setminus \operatorname{dom} f$. Because $A_k \neq \varnothing$, there exists $a_k \in A_k$. Define g by

$$g(j) = \begin{cases} a_k & \text{if } j = k, \\ f(j) & \text{if } j \in \operatorname{dom} f. \end{cases}$$

Then $g \in \mathcal{E}$ and $g > f$, a contradiction to maximality of f.   ∎

4.23 Remarks.   The axiom (ZL) (or more usually (ZL)') is used to assert the existence of an object which cannot be directly constructed, such as

(a) a maximal linearly independent subset in a vector space $V \neq \{0\}$,

(b) a maximal ideal in a ring R,

(c) the choice function sought in the last part of the preceding theorem.

   Proofs involving (ZL)' have a distinct sameness. Let X be an object whose existence we wish to establish. We proceed as follows:

(i) take a non-empty family $\mathcal{E}$ of sets ordered by inclusion, in which X is a (hypothetical) maximal element;

(ii) check that (ZL)' is applicable;

(iii) verify that the maximal element supplied by (ZL)' has all the properties demanded of X.

   Let us review these steps in turn. Choosing $\mathcal{E}$ is usually straightforward. For example, we take $\mathcal{E}$ to be all linearly independent subsets of V in (a) and all proper ideals of R in (b). We then have to exhibit an element of $\mathcal{E}$ to ensure $\mathcal{E} \neq \varnothing$. Again this is easy in our examples: take $\{v\}$, where $0 \neq v \in V$ in (a), and the ideal $\{0\}$ in (b). Notice that $\mathcal{E}$ may be thought of as a family of partial objects, having some of the features X should have.

   Now consider (ii). To confirm that (ZL)' applies, we need to show that the union of a non-empty chain of sets in $\mathcal{E}$ is itself in $\mathcal{E}$. Observe the similarity between the arguments in the proof of (ZL)'' $\Rightarrow$ (KL) and of Lemma 3.4. A chain is a special case of a directed set. In an algebraic $\bigcap$-structure $\mathfrak{L}$ we have $\bigcup \mathcal{D} \in \mathfrak{L}$ whenever $\mathcal{D}$ is a directed subset of $\mathfrak{L}$. In each of our examples above, and in many other (ZL) applications, $\mathcal{E}$ is an algebraic $\bigcap$-structure, and it is this fact that ensures success in (ii).

Step (iii) is immediate in examples (a) and (b), but not in (c). In cases where (iii) is non-trivial, argument by contradiction is invariably used. Exercise 4.20 and the proof of Theorem 9.13 provide illustrations.

Exercises

4.1 For each of the following recursive specifications, construct a map $\Phi\colon P \to P$ whose fixpoints satisfy the specification, show that Φ is order-preserving and describe $\Phi^0(\bot)$, $\Phi^1(\bot)$, $\Phi^2(\bot)$, $\Phi^3(\bot)$, $\Phi^n(\bot)$ and $\bigsqcup_{n \geqslant 0} \Phi^n(\bot)$ via the graphs of the corresponding partial maps or, if you can, via non-recursively defined (partial) maps. The domain of any partial map should be stated explicitly.

(i) $P = (\mathbb{N} \to \mathbb{N}_\bot)$,

$$f(k) = \begin{cases} 1 & \text{if } k = 1, \\ (2k-1) + f(k-1) & \text{otherwise.} \end{cases}$$

(ii) $P = (\mathbb{N}_0 \to (\mathbb{N}_0)_\bot)$,

$$f(k) = \begin{cases} 1 & \text{if } k = 0, \\ f(k+1) & \text{otherwise.} \end{cases}$$

(iii) $P = (\mathbb{N}_0 \times \mathbb{N}_0 \to (\mathbb{N}_0)_\bot)$,

$$f(j, k) = \begin{cases} k & \text{if } j = 0, \\ 1 + f(j-1, k) & \text{otherwise.} \end{cases}$$

(iv) $P = (\mathbb{N}_0 \to (\mathbb{N}_0)_\bot)$,

$$f(k) = \begin{cases} k - 10 & \text{if } k > 100, \\ f(f(k+11)) & \text{otherwise.} \end{cases}$$

4.2 Find all fixpoints of $\Phi\colon (\mathbb{N}_0 \to (\mathbb{N}_0)_\bot) \to (\mathbb{N}_0 \to (\mathbb{N}_0)_\bot)$, where

$$(\Phi(f))(k) = \begin{cases} 1 & \text{if } k = 0, \\ f(k+1) & \text{otherwise.} \end{cases}$$

4.3 Let $P = (\mathbb{Z} \times \mathbb{Z} \to \mathbb{Z}_\bot)$. Let $f \in P$ satisfy

$$f(j, k) = \begin{cases} k + 1 & \text{if } j = k, \\ f(j, f(j-1, k+1)) & \text{otherwise .} \end{cases}$$

Prove that each of

$$(j, k) \mapsto k + 1 \text{ and } (j, k) \mapsto \begin{cases} j + 1 & \text{if } j \geqslant k, \\ k - 1 & \text{otherwise,} \end{cases}$$

is a solution of the above equation. What is the least solution?

4.4   Let $P = (\mathbb{N}_0 \to (\mathbb{N}_0)_\perp)$. Let $f \in P$ satisfy

$$f(k) = \begin{cases} 1 & \text{if } k = 0, \\ f(k+2) & \text{if } k = 1, \\ f(k-2) & \text{otherwise.} \end{cases}$$

Prove that each of

$$k \mapsto 1,$$
$$k \mapsto k \pmod 2 + 1,$$
$$k \mapsto 1 \text{ if } k \text{ is even and } k \mapsto \perp \text{ otherwise,}$$

is a solution of the above equation. What is the least solution?

4.5   Construct the least solutions to the following equations in the CPO
$P = \wp(X)$:

   (i) $S = S \cup T$, where $T \subseteq X$ is fixed;
   (ii) (with $X = \mathbb{N}$) $S = S \cup \{1\} \cup \{n+2 \mid n \in S\}$;
   (iii) (with $X = \mathbb{N}^2$)

$$S = \{(n,n) \mid n \in \mathbb{N}\} \cup \{(n, m+1) \mid (n, m) \in S\};$$

   (iv) (with X a finite group) $S = S \cup \{g_1 g_2 \mid g_1, g_2 \in S\}$.

   [Hint. Find a suitable order-preserving map Φ of which the required set is to be the least fixpoint, guess a formula for $\Phi^n(\perp)$ and
   prove by induction that it works, and finally verify that $\bigsqcup \Phi^n(\perp)$
   is a fixpoint.] (See Theorem 4.5(i).)

4.6   Given $f \colon \mathbb{N}_0 \to (\mathbb{N}_0)_\perp$, let $\overline{f}$ be defined by

$$\overline{f}(k) = \begin{cases} 1 & \text{if } k = 0, \\ k f(k-1) & \text{if } k \geqslant 1 \text{ and } f(k-1) \neq \perp, \\ \perp & \text{otherwise.} \end{cases}$$

Define $\Phi \colon (\mathbb{N}_0 \to (\mathbb{N}_0)_\perp) \to (\mathbb{N}_0 \to (\mathbb{N}_0)_\perp)$ by $\Phi(f) = \overline{f}$.

   (i)   Show that Φ is order-preserving.
   (ii) Prove that Φ is continuous. [Hint. Let $D = \{g_i\}_{i \in I}$ be a
         directed set (or a chain) in $(\mathbb{N}_0 \to (\mathbb{N}_0)_\perp)$, and let $g :=
         \bigsqcup_{i \in I} g_i$. Check that $\overline{g}(k) = \bigsqcup_{i \in I} \overline{g}_i(k)$ for all $k \in \mathbb{N}_0$ by
         considering the cases

(a) $k = 0$,

(b) $g_i(k - 1) = \perp$ for all $i \in I$,

(c) $g_i(k - 1) \neq \perp$ for some $i \in I$.]

(This exercise establishes the continuity of the map of which the factorial function is the least fixpoint.)

4.7 Let P be a CPO. Let $\mu: [P \to P] \to P$ be the fixpoint operator which assigns to each continuous map $f: P \to P$ its least fixpoint $\mu(f)$ and, for each $n \in \mathbb{N}$, let μ_n be defined by $\mu_n(f) = f^n(\perp)$.

(i) Show, by induction, that μ_n is continuous for each n.

(ii) Let $Q = ([P \to P] \to P)$, the CPO of all maps from $[P \to P]$ to P. Show that $\{\mu_n\}_{n \geqslant 1}$ is a chain in Q.

(iii) Show that $\bigsqcup_{n \geqslant 1} \mu_n = \mu$ in Q. Deduce that μ is continuous.

4.8 Consider the following domain equations. In each case, let $\mathbf{F}$ be the corresponding domain constructor. Draw diagrams for $\mathbf{F}^n(\mathbf{1})$ for $n = 1, 2, 3, 4$ and for the least solution P of the equation. Indicate the order-embedding of $\mathbf{F}^n(\mathbf{1})$ into $\mathbf{F}^{n+1}(\mathbf{1})$ for $n = 1, 2, 3$ and of $\mathbf{F}^4(\mathbf{1})$ into P. [Either (a) work abstractly, treating $\mathbf{1}$ as an abstract ordered set, or (b) work concretely, replacing $\mathbf{1}$ by $\mathfrak{N}$ and the operators $\oplus_\vee$, $\oplus_\perp$ and $\oplus$ by $\boxplus_\vee$, $\boxplus_\perp$ and $\boxplus$, respectively. While the abstract approach has the advantage that $\mathbf{F}^n(\mathbf{1})$ is easily found, the embeddings can be more elusive. Using the concrete approach, constructing $\mathbf{F}^n(\mathfrak{N})$ can be a notational nightmare but the embeddings are given at once via Exercise 3.29. In order to solve (v) and (vi) concretely, the operator $\boxplus$ must first be defined— see Exercise 3.30 for a particular case.]

(i) $P \cong \mathbf{1} \oplus_\vee P$.

(ii) $P \cong \mathbf{1} \oplus_\perp P$.

(iii) $P \cong P \oplus \mathbf{1}$. (See Exercise 3.30.)

(iv) $P \cong \mathbf{1} \oplus P \oplus \mathbf{1} = (P \oplus \mathbf{1})_\perp$.

(v) $P \cong P \oplus (\mathbf{1} \oplus_\perp \mathbf{1})$.

(vi) $P \cong (\mathbf{1} \oplus_\perp \mathbf{1}) \oplus P$.

(vii) $P \cong (\mathbf{1} \oplus_\perp P)_\perp$.

4.9 Show that $(\mathbb{N} \to \mathbf{2})$ is the least solution to the domain equation $P \cong P \times \mathbf{2}$.

4.10 Let L be a complete lattice and define $\Phi : \wp(L) \to \wp(L)$ by $\Phi: A \mapsto {\downarrow}\bigvee A$. Show that Φ is order-preserving and that $\mathrm{fix}(\Phi) \cong$

L. (Thus every complete lattice is, up to isomorphism, a lattice of fixpoints.)

4.11 Let P be an ordered set and Q a complete lattice. Given a map $f\colon P \to Q$, define $\overline{f}\colon P \to Q$ by

$$\overline{f}(x) = \bigvee\{\, f(y) \mid y \leqslant x \,\}.$$

Show that $\overline{f}$ is order-preserving and that $f = \overline{f}$ if and only if f is order-preserving. Show further that Φ defined by $\Phi\colon f \mapsto \overline{f}$ is an order-preserving map from Q^P to $Q^{\langle P \rangle} \subseteq Q^P$ whose set of fixpoints is exactly $Q^{\langle P \rangle}$.

4.12   (i) Use the Knaster–Tarski Fixpoint Theorem to prove Banach's Decomposition Theorem:

>   Let X and Y be sets and let $f\colon X \to Y$ and $g\colon Y \to X$ be maps. Then there exist disjoint subsets X_1 and X_2 of X and disjoint subsets Y_1 and Y_2 of Y such that $f(X_1) = Y_1$, $g(Y_2) = X_2$, $X = X_1 \cup X_2$ and $Y = Y_1 \cup Y_2$.
>
>   [Hint. Consider the map $\Phi : \wp(X) \to \wp(X)$ defined by $\Phi(S) = X \smallsetminus g(Y \smallsetminus f(S))$ for $S \subseteq X$.]

    (ii) Use (i) to obtain the Schröder-Bernstein Theorem:

>   Let X and Y be sets and suppose there exist one-to-one maps $f\colon X \to Y$ and $g\colon Y \to X$. Then there exists a bijective map h from X onto Y.

4.13 (For those who know about ordinals.) Let P be a CPO and $\Phi\colon P \to P$ an order-preserving map. Let

$$\Phi^0(\bot) = \bot,$$
$$\Phi^{\beta+1}(\bot) = \Phi(\Phi^\beta(\bot)),$$
$$\Phi^\alpha(\bot) = \bigsqcup\{\, \Phi^\beta(\bot) \mid \beta < \alpha \,\} \text{ if } \alpha \text{ is a limit ordinal.}$$

Prove that $\Phi^\alpha(\bot)$ is well defined for all ordinals α by showing (by transfinite induction) that, if $\alpha \leqslant \beta$ and $\Phi^\beta(\bot)$ is defined, then $\Phi^\alpha(\bot) \leqslant \Phi^\beta(\bot)$. Argue by contradiction to show (with the aid of Hartogs' Theorem) that, for some ordinal α, the point $\Phi^\alpha(\bot)$ is a fixpoint of Φ.

4.14 Let P be a CPO and $\Phi\colon P \to P$ be an order-preserving map. Assume $y \in P$ is such that $\Phi(y) \leqslant y$. Prove that Φ maps ${\downarrow}y$ to ${\downarrow}y$ and, by restricting Φ to ${\downarrow}y$, deduce that $\mu(\Phi) \leqslant y$.

Let C be a chain in fix(Φ) and let $z := \bigsqcup C$. Prove that Φ maps $\uparrow z$ to $\uparrow z$. Deduce that Ψ, the restriction of Φ to $\uparrow z$, has a least fixpoint, given by $\bigsqcup_{\text{fix}(\Phi)} C$. (Hence fix($\Phi$) is a CPO.)

4.15 Let L be a complete lattice and $\Phi \colon L \to L$ be order-preserving. Let X be a subset of fix(Φ). Define

$$Y = \{\, y \in L \mid (\forall x \in X)\, x \leqslant \Phi(y) \leqslant y \,\}$$

and let $\alpha = \bigwedge_L Y$. Prove that $\alpha = \bigvee_{\text{fix}(\Phi)} X$. Deduce that fix($\Phi$) is a complete lattice.

4.16 Let L be a complete lattice and Φ, Ψ be order-preserving maps from L to L. Prove that if Φ and Ψ have a common fixpoint then they have a least common fixpoint, given by

$$\bigwedge \{\, x \in L \mid \Phi(x) \leqslant x \;\&\; \Psi(x) \leqslant x \,\}.$$

Prove further that, if $\Phi \circ \Psi = \Psi \circ \Phi$, then the least common fixpoint of Φ and Ψ is $\mu(\Phi \circ \Psi)$.

4.17 Let P be a CPO and let Φ and Ψ be continuous maps from P to P such that $\Phi \circ \Psi = \Psi \circ \Phi$.

  (i) Show that $\{\, \Phi^m(\Psi^n(\bot)) \mid n, m \geqslant 0 \,\}$ forms a directed set.

  (ii) Show that $\mu(\Phi \circ \Psi)$ is a fixpoint of both Φ and Ψ.

  (iii) Show that $\mu(\Phi \circ \Psi) = \mu(\Phi)$ if and only if $\mu(\Psi) \leqslant \mu(\Phi)$.

4.18 Let L be a complete lattice and suppose that $\Phi \colon L \times L \to L$ and $\Psi \colon L \times L \to L$ are order-preserving maps. Define a map $\Theta \colon L \times L \to L \times L$ by $\Theta(x, y) = (\Phi(x, y), \Psi(x, y))$. Prove that Θ is order-preserving. Prove that the least fixpoint of Θ is (x_0, y_0), where x_0 is the least fixpoint of the map $x \mapsto \Phi(x, \mu(\Psi^x))$ and y_0 is the least fixpoint of the map $y \mapsto \Psi(\mu(\Phi_y), y)$. (Here $\Phi_y \colon x \mapsto \Phi(x, y)$ and $\Psi^x \colon y \mapsto \Psi(x, y)$.)

4.19 Recall that the definition of an **ideal** in a lattice L was given in Exercise 2.11. Denote the set of all ideals of L by $\mathcal{I}(L)$. Deduce from (ZL)′ that any ideal J in L with $J \neq L$ is contained in an ideal I which is maximal in $\langle \mathcal{I}(L) \smallsetminus \{L\}; \subseteq \rangle$.

4.20 Let $\langle P; \leqslant \rangle$ be an ordered set. By applying (ZL)′ to an appropriate family $\mathcal{E}$ of partial orders (regarding an order relation on P as a subset of $P \times P$) show that $\leqslant$ has a linear extension. (This is Szpilrajn's Theorem; you will need its finite version, given in

Exercise 1.22(ii), to prove that the maximal element of $\mathcal{E}$ supplied by (ZL)$'$ is a chain.)

4.21 Let D be a domain.

    (i) Apply (ZL) to show that, if $a \in D$ is finite and $c \in D$ satisfies $c < a$, then there exists $b \in D$ such that $c \leqslant b \prec a$.

    (ii) Show that for all $x, y \in D$ satisfying $x \leqslant y$ the **interval** $[x, y] := {\uparrow}x \cap {\downarrow}y$ is an algebraic lattice.

    (iii) Show that, if $x < y$ in D, then there exist $a, b \in D$ such that $x \leqslant b \prec a \leqslant y$.

4.22 Let L be an algebraic lattice.

    (i) Show that meet distributes over directed joins in L, that is,

$$x \wedge \bigsqcup \{\, y_i \mid i \in I \,\} = \bigsqcup \{\, x \wedge y_i \mid i \in I \,\}.$$

    (ii) Show that if L is distributive then it satisfies the Join-Infinite Distributive Law, that is,

$$x \wedge \bigvee \{\, y_i \mid i \in I \,\} = \bigvee \{\, x \wedge y_i \mid i \in I \,\}.$$

(This result generalizes (the dual of) Exercise 6.10.)

4.23 An element a of a complete lattice is called **completely meet-irreducible** if $a \neq 1$ and $a = \bigwedge S$ implies that $a \in S$ for every subset S of L. Show that in a algebraic lattice L the completely meet-irreducible elements are meet-dense. [Hint. By the dual of Lemma 2.35, it suffices to show that if $s, t \in L$ with $t > s$ then there exists a completely meet-irreducible element m with $m \geqslant s$ and $m \not\geqslant t$. Apply (ZL) to the set $P = \{\, a \in L \mid a \geqslant s \text{ and } a \not\geqslant t \,\}$ then take m to be any maximal element of P.]

5

Lattices as Algebraic Structures

In Chapter 2 lattices were introduced as ordered sets of a special type. Here we consider them as algebraic structures in their own right, in a way that is reminiscent of the study of, for example, groups or rings.

Lattices as algebraic structures

It was noted in 2.5 that, given a lattice L, we may define binary operations **join** and **meet** on the non-empty set L by

$$a \vee b := \sup\{a, b\} \quad \text{and} \quad a \wedge b := \inf\{a, b\} \quad (a, b \in L).$$

In this section we explore the algebraic properties of these operations. We first amplify the connection between $\vee, \wedge$ and $\leqslant$. Since we shall often use the following lemma it deserves a name.

5.1 The Connecting Lemma. *Let L be a lattice and let $a, b \in L$. Then the following are equivalent:*

(i) $a \leqslant b$;

(ii) $a \vee b = b$;

(iii) $a \wedge b = a$.

Proof. It was shown in 2.4(1) that (i) implies both (ii) and (iii). Now assume (ii). Then b is an upper bound for $\{a, b\}$, whence $b \geqslant a$. Thus (i) holds. By duality, (iii) implies (i) also. ∎

5.2 Theorem. *Let L be a lattice. Then $\vee$ and $\wedge$ satisfy, for all $a, b, c \in L$,*

(L1)	$(a \vee b) \vee c = a \vee (b \vee c)$	*(associative laws)*
(L1)$^\partial$	$(a \wedge b) \wedge c = a \wedge (b \wedge c)$	
(L2)	$a \vee b = b \vee a$	*(commutative laws)*
(L2)$^\partial$	$a \wedge b = b \wedge a$	
(L3)	$a \vee a = a$	*(idempotency laws)*
(L3)$^\partial$	$a \wedge a = a$	
(L4)	$a \vee (a \wedge b) = a$	*(absorption laws)*
(L4)$^\partial$	$a \wedge (a \vee b) = a.$	

Proof. Note that the dual of a statement about lattices phrased in terms of $\vee$ and $\wedge$ is obtained simply by interchanging $\vee$ and $\wedge$ (this is the **Duality Principle for Lattices**). It is therefore enough to consider (L1)–(L4).

We proved (L3) in 2.4(1) and (L2) is immediate because, for any set S, $\sup S$ is independent of the order in which the elements of S are listed. Also, (L4) follows easily from the Connecting Lemma. To prove (L1) it is enough, thanks to (L2), to show that $(a \vee b) \vee c = \sup\{a, b, c\}$. This is the case if $\{(a \vee b), c\}^u = \{a, b, c\}^u$. But

$$d \in \{a, b, c\}^u \iff d \in \{a, b\}^u \text{ and } d \geqslant c$$
$$\iff d \geqslant a \vee b \text{ and } d \geqslant c$$
$$\iff d \in \{(a \vee b), c\}^u. \qquad \blacksquare$$

We now turn things round and start from a set carrying operations $\vee$ and $\wedge$ which satisfy the identities given in the preceding theorem.

5.3 Theorem. *Let $\langle L; \vee, \wedge \rangle$ be a non-empty set equipped with two binary operations which satisfy (L1)–(L4) and $(L1)^{\partial}$–$(L4)^{\partial}$ from 5.2.*

(i) *For all $a, b \in L$, $a \vee b = b$ if and only if $a \wedge b = a$.*

(ii) *Define $\leqslant$ by $a \leqslant b$ if $a \vee b = b$. Then $\leqslant$ is an order relation.*

(iii) *With $\leqslant$ as in (ii), $\langle L; \leqslant \rangle$ is a lattice in which the original operations agree with the induced operations, that is, for all $a, b \in L$,*

$$a \vee b = \sup\{a, b\} \text{ and } a \wedge b = \inf\{a, b\}.$$

Proof. Assume $a \vee b = b$. Then

$$a = a \wedge (a \vee b) \qquad \text{(by } (L4)^{\partial})$$
$$= a \wedge b \qquad \text{(by assumption).}$$

Conversely, assume $a \wedge b = a$. Then

$$b = b \vee (b \wedge a) \qquad \text{(by (L4))}$$
$$= b \vee (a \wedge b) \qquad \text{(by } (L2)^{\partial})$$
$$= b \vee a \qquad \text{(by assumption)}$$
$$= a \vee b \qquad \text{(by (L2)).}$$

Now define $\leqslant$ as in (ii). Then $\leqslant$ is reflexive by (L3), antisymmetric by (L2) and transitive by (L1). The details are left as an exercise.

To show that, in the ordered set $\langle L; \leqslant \rangle$, $\sup\{a, b\} = a \vee b$ it must first be checked that $a \vee b \in \{a, b\}^u$ and second that $d \in \{a, b\}^u$ implies $d \geqslant a \vee b$. To do this, remember that $\leqslant$ is given by $p \leqslant q$ if and only if $p \vee q = q$ and justify each step by appealing to one of the identities. The characterization of inf is obtained by duality (again). $\qquad \blacksquare$

5.4 Stocktaking. We have shown that lattices can be completely characterized in terms of the join and meet operations. We may henceforth say 'let L be a lattice', replacing L by $\langle L; \leqslant \rangle$ or by $\langle L; \vee, \wedge \rangle$ if we want to emphasize that we are thinking of it as a special kind of ordered set or as an algebraic structure.

5.5 Remark. Let L be a lattice and let $a, b, c \in L$ and assume that $b \leqslant a \leqslant b \vee c$. Then

$$b \vee c \leqslant a \vee c \leqslant (b \vee c) \vee c = b \vee c,$$

whence $a \vee c = b \vee c$; see Figure 5.1. This simple observation and its dual are particularly useful when calculating joins and meets on a diagram— once we know the join of b and c, the join of c with the intermediate element a is forced.

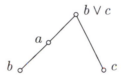

Figure 5.1

5.6 Definitions. Let L be a lattice. It may happen that $\langle L; \leqslant \rangle$ has top and bottom elements $\top$ and $\bot$ as defined in 1.23. When thinking of L as $\langle L; \vee, \wedge \rangle$ it is appropriate to view these elements from a more algebraic standpoint. We say L has a **unit**, or **identity**, element if there exists $1 \in L$ such that $a = a \wedge 1$ for all $a \in L$. Dually, L is said to have a **zero** element if there exists $0 \in L$ such that $a = a \vee 0$ for all $a \in L$. The lattice $\langle L; \vee, \wedge \rangle$ has a unit element 1 if and only if $\langle L; \leqslant \rangle$ has a top element $\top$ and, in that case, $1 = \top$. A dual statement holds for 0 and $\bot$. A lattice $\langle L; \vee, \wedge \rangle$ possessing 0 and 1 is called **bounded**. A finite lattice is automatically bounded, with $1 = \bigvee L$ and $0 = \bigwedge L$. Recalling 2.7(4), note that $\langle \mathbb{N}_0; \mathrm{lcm}, \gcd \rangle$ is bounded—with $1 = 0$ and $0 = 1$!!

Sublattices, products and homomorphic images

This section presents methods for deriving new lattices.

5.7 Sublattices. Let L be a lattice and $\varnothing \neq M \subseteq L$. Then M is a **sublattice** of L if

$$a, b \in M \text{ implies } a \vee b \in M \text{ and } a \wedge b \in M.$$

We denote the collection of all sublattices of L by $\operatorname{Sub} L$ and let $\operatorname{Sub}_0 L = \operatorname{Sub} L \cup \{\varnothing\}$. Then $\operatorname{Sub}_0 L$ is a topped $\bigcap$-structure on L and so a complete lattice.

5.8 Examples.

(1) Any one-element subset of a lattice is a sublattice. More generally, any chain in a lattice is a sublattice. (In fact, when testing that a non-empty subset M is a sublattice it is sufficient to consider non-comparable elements a, b in 5.7.)

(2) In the diagrams in Figure 5.2 the shaded elements in lattices (i) and (ii) form sublattices, while those in (iii) and (iv) do not.

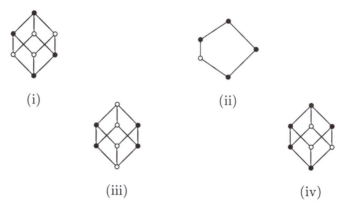

(i)   (ii)

(iii)   (iv)

Figure 5.2

(3) A subset M of a lattice $\langle L; \leqslant \rangle$ may be a lattice in its own right without being a sublattice of L; see Figure 5.2(iv) for an example.

5.9 Products.

Let L and K be lattices. Define $\vee$ and $\wedge$ coordinate-wise on $L \times K$, as follows:

$$(\ell_1, k_1) \vee (\ell_2, k_2) = (\ell_1 \vee \ell_2, k_1 \vee k_2),$$
$$(\ell_1, k_1) \wedge (\ell_2, k_2) = (\ell_1 \wedge \ell_2, k_1 \wedge k_2).$$

It is routine to check that $L \times K$ is a lattice. Also

$$(\ell_1, k_1) \vee (\ell_2, k_2) = (\ell_2, k_2) \iff \ell_1 \vee \ell_2 = \ell_2 \text{ and } k_1 \vee k_2 = k_2$$
$$\iff \ell_1 \leqslant \ell_2 \text{ and } k_1 \leqslant k_2$$
$$\iff (\ell_1, k_1) \leqslant (\ell_2, k_2),$$

with respect to the order on $L \times K$ defined in 1.26. Hence the lattice formed by taking the ordered set product of lattices L and K is the same as that obtained by defining $\vee$ and $\wedge$ coordinatewise on $L \times K$.

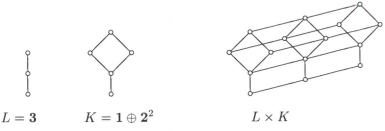

Figure 5.3

Figure 5.3 shows the product of the lattices $L = \mathbf{3}$ and $K = \mathbf{1} \oplus \mathbf{2}^2$. Notice how (isomorphic copies) of L and K sit inside $L \times K$ as the sublattices $L \times \{0\}$ and $\{0\} \times K$. It is routine to show that the product of lattices L and K always contains sublattices isomorphic to L and K.

Iterated products and powers are defined in the obvious way. It is possible to define the product of an infinite family of lattices, but we shall not need this construction.

5.10 Homomorphisms. From the viewpoint of this chapter it is natural to regard as canonical those maps between lattices which preserve the operations join and meet. Since lattices are also ordered sets, order-preserving maps are also available. We need to explore the relationship between these classes of maps. We begin with some definitions.

Let L and K be lattices. A map $f\colon L \to K$ is said to be a **homomorphism** (or, for emphasis, **lattice homomorphism**) if f is **join-preserving** and **meet-preserving**, that is, for all $a, b \in L$,

$$f(a \vee b) = f(a) \vee f(b) \quad \text{and} \quad f(a \wedge b) = f(a) \wedge f(b).$$

A bijective homomorphism is a **(lattice-)isomorphism**. If $f\colon L \to K$ is a one-to-one homomorphism, then the sublattice $f(L)$ of K is isomorphic to L and we refer to f as an **embedding** (of L **into** K). For bounded lattices L and K it is often appropriate to consider homomorphisms $f\colon L \to K$ such that $f(0) = 0$ and $f(1) = 1$. Such maps are called $\{0,1\}$-**homomorphisms**. They arise in Chapters 8 and 10.

Referring back to Figure 1.3, we see that each of φ_2–φ_6 is an order-preserving map from one lattice to another. The maps φ_2 and φ_3 are homomorphisms, the remainder are not. Neither join nor meet is preserved by φ_4. The map φ_5 preserves joins but does not preserve all meets; φ_6 is meet-preserving but does not preserve all joins. Thus in general an order-preserving map may not be a homomorphism. Conveniently, a possible demarcation dispute between order-isomorphism and lattice-isomorphism does not arise, as 5.11(ii) shows.

5.11 Proposition. *Let L and K be lattices and $f: L \to K$ a map.*

(i) *The following are equivalent:*

   (a) *f is order-preserving;*

   (b) *$(\forall a, b \in L)\, f(a \vee b) \geqslant f(a) \vee f(b)$;*

   (c) *$(\forall a, b \in L)\, f(a \wedge b) \leqslant f(a) \wedge f(b)$.*

   In particular, if f is a homomorphism, then f is order-preserving.

(ii) *The following are equivalent:*

   (a) *f is an order-isomorphism;*

   (b) *f is bijective and an order-embedding;*

   (c) *f is a lattice-isomorphism.*

Proof. Part (i) is an easy consequence of the Connecting Lemma. Part (ii) follows from (i) and Exercise 2.10. ∎

Congruences

In group theory courses it is customary, after homomorphisms have been introduced, to go on to define normal subgroups and quotient groups (factor groups) and to reveal the intimate connection between these concepts that is summed up in the Fundamental Homomorphism Theorem. In this section we develop the rudiments of the corresponding theory for lattices.

We begin with a summary of the basic group theory results, expressed in a form that will make the parallels with the lattice case stand out clearly. This summary is prefaced by a brief refresher on equivalence relations.

5.12 Equivalence relations: a recap. We recall that an equivalence relation on a set X is a binary relation on X which is reflexive, symmetric and transitive. We write either $a \equiv b \,(\mathrm{mod}\,\theta)$ or $(a, b) \in \theta$ to indicate that a and b are related under the relation θ. The latter notation is used where it is appropriate to be formally correct and to regard θ as a subset of $X \times X$.

An equivalence relation θ on X gives rise to a partition of X into disjoint subsets. These subsets are the equivalence classes or **blocks** of θ; we shall usually use the term block, as it is more in keeping with the pictorial approach we shall be adopting. A typical block is of the form $[a]_\theta := \{\, x \in X \mid x \equiv a \,(\mathrm{mod}\,\theta)\,\}$. In the opposite direction, a partition of X into a union of non-empty disjoint subsets gives rise to an equivalence relation whose blocks are the subsets in the partition. This correspondence between equivalence relations and partitions is set out more explicitly in, for example, [9].

5.13 The group case. Let G and H be groups and $f : G \to H$ be a group homomorphism. We may define an equivalence relation θ on G by

$$(\forall a, b \in G)\, a \equiv b \,(\mathrm{mod}\,\theta) \iff f(a) = f(b).$$

This relation and the partition of G it induces have the following important properties.

(1) The relation θ is compatible with the group operation in the sense that, for all $a, b, c, d \in G$,

$$a \equiv b \,(\mathrm{mod}\,\theta) \,\&\, c \equiv d \,(\mathrm{mod}\,\theta) \implies ac \equiv bd \,(\mathrm{mod}\,\theta).$$

(2) The block $N = [1]_\theta := \{\, g \in G \mid g \equiv 1 \,(\mathrm{mod}\,\theta)\,\}$ is a normal subgroup of G.

(3) For each $a \in G$, the block $[a]_\theta := \{\, g \in G \mid g \equiv a \,(\mathrm{mod}\,\theta)\,\}$ equals the (left) coset $aN := \{\, an \mid n \in N \,\}$.

(4) The natural definition,

$$[a]_\theta [b]_\theta := [ab]_\theta \quad \text{for all } a, b \in G,$$

yields a well-defined group operation on $\{\, [a]_\theta \mid a \in G \,\}$; by (2) and (3), the resulting group is precisely the quotient group G/N and hence (by the Fundamental Homomorphism Theorem for groups) is isomorphic to the subgroup $f(G)$ of H.

5.14 Speaking universally. It should now be apparent that much of the above is not particular to groups and group homomorphisms, but will apply, *mutatis mutandis*, to lattices and lattice homomorphisms. In fact, the natural setting for the Fundamental Homomorphism Theorem and its consequences is not group theory or lattice theory but **universal algebra**. This is the general theory of classes of algebraic structures, of which groups, rings, lattices, bounded lattices, vector spaces, ... are examples. Lattice theory and universal algebra have a close and symbiotic relationship: results from universal algebra (such as the Fundamental Homomorphism Theorem) specialize to classes of lattices, and lattices arise naturally in the study of abstract algebras, as lattices of congruences, for example. For references, see the Appendix.

5.15 Lemma. *Let L and K be lattices and let $f : L \to K$ be a lattice homomorphism. Then the equivalence relation θ defined on L by*

$$(\forall a, b \in L)\, a \equiv b \,(\mathrm{mod}\,\theta) \iff f(a) = f(b)$$

is compatible with the join and meet operations on L, in the sense that, for all $a, b, c, d \in L$,

$$a \equiv b \,(\mathrm{mod}\,\theta) \text{ and } c \equiv d \,(\mathrm{mod}\,\theta)$$

imply

$$a \vee c \equiv b \vee d \,(\mathrm{mod}\,\theta) \text{ and } a \wedge c \equiv b \wedge d \,(\mathrm{mod}\,\theta).$$

Proof. It is elementary that θ is indeed an equivalence relation. Now assume $a \equiv b \,(\mathrm{mod}\,\theta)$ and $c \equiv d \,(\mathrm{mod}\,\theta)$, so that $f(a) = f(b)$ and $f(c) = f(d)$. Hence, since f preserves join,

$$f(a \vee c) = f(a) \vee f(c) = f(b) \vee f(d) = f(b \vee d).$$

Therefore $a \vee c \equiv b \vee d \,(\mathrm{mod}\,\theta)$. Dually, θ is compatible with meet. ∎

5.16 Definitions and examples. An equivalence relation on a lattice L which is compatible with both join and meet is called a **congruence** on L. If L and K are lattices and $f : L \to K$ is a lattice homomorphism, then the associated congruence θ on L defined in 5.15 is known as the **kernel** of f and is denoted by $\ker f$. The set of all congruences on L is denoted by $\mathrm{Con}\, L$. Examples of homomorphisms and their kernels are given in Figure 5.4.

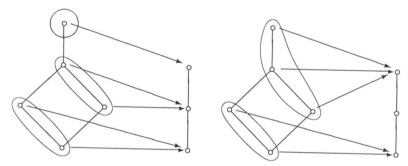

Figure 5.4

Note how a congruence on a lattice L can be indicated on a diagram by placing a loop around the elements in each block of the corresponding partition.

The following lemma is handy when calculating with congruences.

5.17 Lemma. (i) *An equivalence relation θ on a lattice L is a congruence if and only if, for all $a, b, c \in L$,*

$$a \equiv b \,(\mathrm{mod}\,\theta) \implies a \vee c \equiv b \vee c \,(\mathrm{mod}\,\theta) \text{ and } a \wedge c \equiv b \wedge c \,(\mathrm{mod}\,\theta).$$

(ii) *Let θ be a congruence on L and let $a, b, c \in L$.*

(a) If $a \equiv b \,(\mathrm{mod}\,\theta)$ and $a \leqslant c \leqslant b$, then $a \equiv c \,(\mathrm{mod}\,\theta)$.

(b) $a \equiv b \,(\mathrm{mod}\,\theta)$ if and only if $a \wedge b \equiv a \vee b \,(\mathrm{mod}\,\theta)$.

Proof. (i) Assume that θ is a congruence on L. If $a \equiv b \,(\mathrm{mod}\,\theta)$, then, since $c \equiv c \,(\mathrm{mod}\,\theta)$, we have $a \vee c \equiv b \vee c \,(\mathrm{mod}\,\theta)$ and $a \wedge c \equiv b \wedge c \,(\mathrm{mod}\,\theta)$. The converse is left as an exercise.

(ii) Let θ be a congruence on L. To prove (a), note first that $a \leqslant c \leqslant b$ implies $a = a \wedge c$ and $c = b \wedge c$. Assume $a \equiv b \,(\mathrm{mod}\,\theta)$. Then $a \wedge c \equiv b \wedge c \,(\mathrm{mod}\,\theta)$, so $a \equiv c \,(\mathrm{mod}\,\theta)$.

Finally we consider (b). If $a \equiv b \,(\mathrm{mod}\,\theta)$, then $a \vee a \equiv b \vee a \,(\mathrm{mod}\,\theta)$ and $a \wedge a \equiv b \wedge a \,(\mathrm{mod}\,\theta)$ by the definition of a congruence. The lattice identities imply $a \equiv a \vee b \,(\mathrm{mod}\,\theta)$ and $a \equiv a \wedge b \,(\mathrm{mod}\,\theta)$. Since θ is transitive and symmetric, we deduce $a \wedge b \equiv a \vee b \,(\mathrm{mod}\,\theta)$.

Conversely, assume $a \wedge b \equiv a \vee b \,(\mathrm{mod}\,\theta)$. We have $a \wedge b \leqslant a \leqslant a \vee b$, so $a \wedge b \equiv a \,(\mathrm{mod}\,\theta)$, by (a), and similarly $a \wedge b \equiv b \,(\mathrm{mod}\,\theta)$. Because θ is symmetric and transitive, it follows that $a \equiv b \,(\mathrm{mod}\,\theta)$. ∎

5.18 The lattice of congruences. An equivalence relation θ on a lattice L is a subset of L^2 (see 5.12). We can rewrite the compatibility conditions in the form

$$(a, b) \in \theta \text{ and } (c, d) \in \theta \text{ imply } (a \vee c, b \vee d) \in \theta \text{ and } (a \wedge c, b \wedge d) \in \theta.$$

As this says precisely that θ is a sublattice of L^2 we could define congruences to be those subsets of L^2 which are both equivalence relations and sublattices of L^2. With this viewpoint, the set $\mathrm{Con}\, L$ of congruences on a lattice L is a family of sets. It is easily seen to be a topped $\bigcap$–structure on L^2. Hence $\mathrm{Con}\, L$, when ordered by inclusion, is a complete lattice, by 2.17. The least element, **0**, and greatest element, **1**, are given by $\mathbf{0} = \{ (a, a) \mid a \in L \}$ and $\mathbf{1} = L^2$.

5.19 Quotient lattices. Given an equivalence relation θ on a lattice L there is a natural way to try to define operations $\vee$ and $\wedge$ on the set

$$L/\theta := \{ [a]_\theta \mid a \in L \}$$

of blocks. Namely, for all $a, b \in L$, we 'define'

$$[a]_\theta \vee [b]_\theta := [a \vee b]_\theta \text{ and } [a]_\theta \wedge [b]_\theta := [a \wedge b]_\theta.$$

These will produce well-defined operations precisely when the definitions are independent of the elements chosen to represent the equivalence classes, that is, when

$$[a_1]_\theta = [a_2]_\theta \text{ and } [b_1]_\theta = [b_2]_\theta$$

imply
$$[a_1 \vee b_1]_\theta = [a_2 \vee b_2]_\theta \text{ and } [a_1 \wedge b_1]_\theta = [a_2 \wedge b_2]_\theta,$$
for all $a, b \in L$. Since, for all $a_1, a_2 \in L$,
$$[a_1]_\theta = [a_2]_\theta \Leftrightarrow a_1 \in [a_2]_\theta \Leftrightarrow (a_1, a_2) \in \theta \Leftrightarrow a_1 \equiv a_2 \,(\mathrm{mod}\,\theta),$$
it follows that $\vee$ and $\wedge$ are well defined on L/θ if and only if θ is a congruence. When θ is a congruence on L, we call $\langle L/\theta; \vee, \wedge \rangle$ the **quotient lattice of L modulo θ**. Our next lemma justifies this terminology. It is proved with the aid of 5.15.

5.20 Lemma. *Let θ be a congruence on the lattice L. Then $\langle L/\theta; \vee, \wedge \rangle$ is a lattice and the natural quotient map $q \colon L \to L/\theta$, defined by $q(a) := [a]_\theta$, is a homomorphism.*

We can now state the Fundamental Homomorphism Theorem for lattices. Its proof is a routine verification.

5.21 Theorem. *Let L and K be lattices, let f be a homomorphism of L onto K and define $\theta = \ker f$. Then the map $g \colon L/\theta \to K$, given by $g([a]_\theta) = f(a)$ for all $[a]_\theta \in L/\theta$, is well defined, that is,*
$$(\forall a, b \in L)\, [a]_\theta = [b]_\theta \text{ implies } g([a]_\theta) = g([b]_\theta).$$

Moreover g is an isomorphism between L/θ and K. Furthermore, if q denotes the quotient map, then $\ker q = \theta$ and the diagram in Figure 5.5 commutes, that is, $g \circ q = f$.

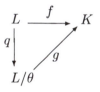

Figure 5.5

In our discussion of congruences we have so far treated lattices as algebraic structures; the underlying order on a lattice has not been mentioned. In the remainder of this section we redress the balance.

5.22 Congruences and diagrams.   Some examples of congruences and the resulting quotient lattices are given in Figure 5.6.

When considering the blocks of a congruence θ on L it is best to think of each block A as an entity in its own right rather than as the block $[a]_\theta$ asssociated with some $a \in L$, as the latter gives undue emphasis to the element a. Intuitively, the quotient lattice L/θ is obtained by collapsing each block to a point.

Assume we are given a diagram of a finite lattice L and loops are drawn on the diagram representing a partition of L. Two natural geometric questions arise.

(a) How can we tell if the equivalence relation corresponding to the partition is a congruence ?

(b) If we know that the loops define the blocks of a congruence θ, how do we go about drawing L/θ?

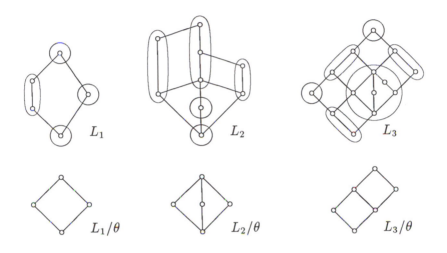

Figure 5.6

There is, of course, no definitive way to draw a lattice diagram. By providing a description of the order and the covering relation on L/θ, Lemma 5.23 provides as good an answer as we can expect to Question (b). The proof of the lemma is left as an easy exercise.

5.23 Lemma. *Let θ be a congruence on a lattice L and let A and B be blocks of θ. Then*

(i) *$A \leqslant B$ in L/θ if and only if there exist $a \in A$ and $b \in B$ such that $a \leqslant b$.*

(ii) *$A \prec B$ in L/θ if and only if $A < B$ in L/θ and $a \leqslant c \leqslant b$ implies $c \in A$ or $c \in B$, for all $a \in A$, all $b \in B$ and all $c \in L$.*

(iii) *If $a \in A$ and $b \in B$ then $a \vee b \in A \vee B$ and $a \wedge b \in A \wedge B$.*

We pursue an answer to Question (a) by first looking for properties that the blocks of a congruence must possess. The blocks are certainly sublattices and are convex, in an obvious order-theoretic sense. (A subset Q of an ordered set P is **convex** if $x \leqslant z \leqslant y$ implies $z \in Q$ whenever $x, y \in Q$ and $z \in P$.) A third property, also with a geometric flavour, relates elements in different blocks. It is described in 5.24. Theorem 5.25 shows that these properties of blocks characterize congruences among equivalence relations.

5.24 The quadrilateral argument.   Let L be a lattice and suppose that $\{a, b, c, d\}$ is a four-element subset of L. Then a, b and c, d are said to be opposite sides of the **quadrilateral** $\langle a, b; c, d \rangle$ if $a < b$, $c < d$ and either

$$(a \vee d = b \text{ and } a \wedge d = c) \quad \text{or} \quad (b \vee c = d \text{ and } b \wedge c = a).$$

We say that the blocks of a partition of L are **quadrilateral-closed** if whenever a, b and c, d are opposite sides of a quadrilateral and $a, b \in A$ for some block A then $c, d \in B$ for some block B (see Figure 5.7). Note that for a covering pair $a \prec b$, we often indicate $a \equiv b \pmod{\theta}$ on a diagram by drawing a wavy line from a to b, to be thought of as a spring which collapses a and b together. See the $\mathbf{N}_5$ in Figure 5.9.

Figure 5.7

5.25 Theorem.   Let L be a lattice and let θ be an equivalence relation on L. Then θ is a congruence if and only if

(i) *each block of θ is a sublattice of L,*

(ii) *each block of θ is convex,*

(iii) *the blocks of θ are quadrilateral-closed.*

Proof. Although the necessity of (i), (ii) and (iii) follows easily from Lemma 5.17 we elect to give a proof based on Lemma 5.23 as this illustrates the use of blocks of θ rather than the relation θ itself. Assume that θ is a congruence on L and let A and B be blocks of θ.

(i) If $a, b \in A$, then $a \vee b \in A \vee A = A$ and $a \wedge b \in A \wedge A = A$, by 5.23(iii). Hence A is a sublattice of L.

(ii) Let $a, b \in A$, let $c \in L$ with $a \leqslant c \leqslant b$ and assume that c belongs to the block C of θ. Then, by 5.23(i), we have $A \leqslant C \leqslant A$ in L/θ and hence $A = C$. Thus $c \in C = A$ and hence A is convex.

(iii) Let a, b and c, d be opposite sides of a quadrilateral, with $a \vee d = b$ and $a \wedge d = c$ (see Figure 5.7). We assume that $a, b \in A$ and $d \in B$ and seek to prove that $c \in B$. Since $d \leqslant b$ we have $B \leqslant A$ (by 5.23(i)) and thus $c = a \wedge d \in A \wedge B = B$, as required.

The converse is much harder. Assume that (i), (ii) and (iii) hold. By Lemma 5.17, θ is a congruence provided that, for all $a, b, c, \in L$,

$$a \equiv b \,(\text{mod}\,\theta) \text{ implies } a \vee c \equiv b \vee c \,(\text{mod}\,\theta) \text{ and } a \wedge c \equiv b \wedge c \,(\text{mod}\,\theta).$$

Let $a, b, c \in L$ with $a \equiv b \,(\text{mod}\,\theta)$. By duality it is enough to show that $a \vee c \equiv b \vee c \,(\text{mod}\,\theta)$. Define $A := [a]_\theta = [b]_\theta$. Since A is a sublattice of L, we have $x := a \wedge b \in A$ and $y := a \vee b \in A$. Our first claim is that $x \vee c \equiv y \vee c \,(\text{mod}\,\theta)$. We distinguish two cases (see Figure 5.8).

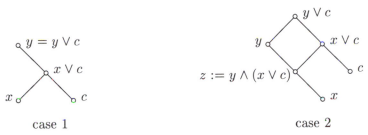

case 1   case 2

Figure 5.8

Case 1: $c \leqslant y$. We have $x \leqslant x \vee c \leqslant y \vee c = y$; the second inequality holds because $x < y$ (see Exercise 2.8) and the final equality follows from the Connecting Lemma. Hence $x \vee c \equiv y \vee c \,(\text{mod}\,\theta)$, since the block A contains both x and y and is convex.

Case 2: $c \not\leqslant y$. Since $x \leqslant y$, we have $x \vee c \leqslant y \vee c$. If $x \vee c = y \vee c$, then $x \vee c \equiv y \vee c \,(\text{mod}\,\theta)$ as θ is reflexive; thus we may assume that $x \vee c < y \vee c$. Since x is a lower bound of $\{y, x \vee c\}$ we have

$$x \leqslant z := y \wedge (x \vee c) \leqslant y.$$

Now $x \leqslant z \leqslant x \vee c$ implies $z \vee c = x \vee c$ (see Exercise 2.8) and thus $z \neq y$ as $y \vee c > x \vee c$. Thus z, y and $x \vee c, y \vee c$ are opposite sides of a quadrilateral. Since the block A is convex and $x, y \in A$, it follows that $z \in A$. Since $z, y \in A$ and θ is quadrilateral-closed it follows that $x \vee c$ and $y \vee c$ belong to the same block, say B. Thus $x \vee c \equiv y \vee c \,(\text{mod}\,\theta)$, as claimed.

We show $a \vee c \equiv b \vee c \pmod{\theta}$ by showing that $a \vee c$ and $b \vee c$ both belong to the block B. Since $a \wedge b \leqslant a \leqslant a \vee b$ and $a \wedge b \leqslant b \leqslant a \vee b$ we have (by Exercise 2.8 again).

$$x \vee c = (a \wedge b) \vee c \leqslant a \vee c \leqslant a \vee b \vee c = y \vee c$$

and

$$x \vee c = (a \wedge b) \vee c \leqslant b \vee c \leqslant a \vee b \vee c = y \vee c.$$

Since $x \vee c, y \vee c \in B$ and B is convex, it follows that $a \vee c, b \vee c \in B$. ∎

5.26 Principal congruences. The characterization in 5.25 allows us to see how a congruence spreads through a lattice L, that is, it permits us to answer the question:

> 'Which pairs of elements c, d in L must be collapsed in order to obtain a congruence θ which collapses a given pair a, b?'

The smallest congruence collapsing a and b is denoted by $\theta(a, b)$ and is called the **principal congruence generated by** (a, b). Since $\operatorname{Con} L$ is an $\bigcap$-structure, $\theta(a, b)$ exists for all $(a, b) \in L^2$: indeed,

$$\theta(a, b) = \bigwedge \{ \theta \in \operatorname{Con} L \mid (a, b) \in \theta \}.$$

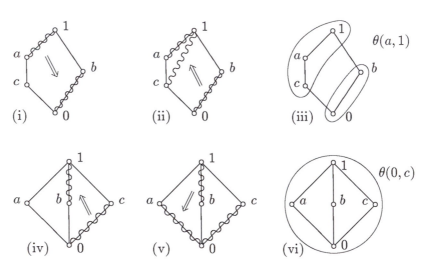

Figure 5.9

5.27 Examples. The diagrams of $\mathbf{N}_5$ and $\mathbf{M}_3$ in Figure 5.9 show the partitions corresponding to the principal congruences $\theta(a, 1)$ and $\theta(0, c)$ respectively. We use the quadrilateral argument to justify these claims.

(1) To find the blocks of the principal congruence $\theta(a,1)$ on $\mathbf{N}_5$, we first use the quadrilateral $\langle a,1;0,b\rangle$ to show that $a \equiv 1$ implies $0 \equiv b$ (here $\equiv$ denotes equivalence with respect to $\theta(a,1)$). The quadrilateral $\langle 0,b;c,1\rangle$ yields $c \equiv 1 \,(\mathrm{mod}\,\theta)$. Since blocks of $\theta(a,1)$ are convex, we deduce that $a,c,1$ lie in the same block. It is clear that $\{0,b\}$ and $\{a,c,1\}$ are convex sublattices and together are quadrilateral-closed. Thus they form the blocks of $\theta(a,1)$ on $\mathbf{N}_5$.

(2) The diagrams in Figure 5.9(iv)–(vi) illustrate the application of the quadrilateral argument to $\mathbf{M}_3$, starting with the pair $(0,c)$. After step (ii) we deduce that $a,c,0$ lie in the same block, say A. Since the blocks of a congruence are sublattices, we have $1 = a \vee c \in A$ and $0 = a \wedge c \in A$. Thus, since blocks are convex, A is the only block. Hence $\theta(0,c) = \mathbf{1}$.

Exercises

Exercises from the text. Complete the proof of Theorem 5.2. Prove that $\mathrm{Sub}_0 L$ is a topped $\bigcap$-structure (see 5.6). Fill in the details of the proof of Proposition 5.11. Complete the proof of 5.17(i). Prove Lemma 5.20, Theorem 5.21 and Lemma 5.23.

5.1   In Theorem 5.2 more identities were listed than was necessary. Prove that each of (L3) and (L3)$^{\partial}$ may be derived from (L4) and (L4)$^{\partial}$. (Do not use any other identities.)

5.2   Let $A = (a_{ij})$ be an $m \times n$ matrix whose entries are elements of a lattice L.

   (i) Prove the Mini-Max Theorem, *viz.*
$$\bigvee_{j=1}^{n}\left(\bigwedge_{i=1}^{m} a_{ij}\right) \leqslant \bigwedge_{k=1}^{m}\left(\bigvee_{\ell=1}^{n} a_{k\ell}\right),$$
   that is, (the join of the meets of the columns of A) $\leqslant$ (the meet of the joins of the rows of A). (Note that the associative laws allow iterated joins or meets to be written unambiguously without brackets.)

   (ii) By applying (i) to a suitable 2×2 matrix, derive the **distributive inequality**
$$a \wedge (b \vee c) \geqslant (a \wedge b) \vee (a \wedge c).$$

   (iii) By applying (i) to a suitable 3×3 matrix, derive the **median inequality**
$$(a \wedge b) \vee (b \wedge c) \vee (c \wedge a) \leqslant (a \vee b) \wedge (b \vee c) \wedge (c \vee a).$$

[Hint. Each row and each column of such a 3×3 matrix must contain exactly two of the elements a, b, c with one repetition.]

5.3 Consider the lattices L_1, L_2 and L_3 in Figure 5.10.

(i) Find L_1 as a sublattice of L_2.

(ii) The shaded elements of L_3 do not form a sublattice. Why?

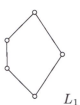 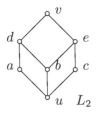 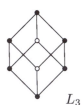

Figure 5.10

5.4 Let L be a lattice. Prove that the following are equivalent:

(i) L is a chain;

(ii) every non-empty subset of L is a sublattice;

(iii) every two-element subset of L is a sublattice.

5.5 Let L be a lattice and for each $X \subseteq L$ let

$$[X] := \bigcap \{ K \in \mathrm{Sub}_0 L \mid X \subseteq K \}.$$

Then $[-] \colon \wp(L) \to \wp(L)$ is the closure operator corresponding to the topped $\bigcap$-structure $\mathrm{Sub}_0 L$; see 2.20 and 2.21.

If X is non-empty, then $[X]$ is the smallest sublattice of L which contains X and is called the **sublattice generated by** X. The definition of $[X]$ in terms of set-intersection does not give a viable method for calculating $[X]$ in a finite lattice. Here is an alternative method for obtaining $[X]$ from a diagram of L. Let $\varnothing \ne X \subseteq L$ and define recursively

$$X_0 := X \quad \text{and} \quad X_{k+1} := \{ a \vee b \mid a, b \in X_k \} \cup \{ a \wedge b \mid a, b \in X_k \}.$$

(i) Show that $X_k \subseteq X_{k+1}$ for all $k \in \mathbb{N}_0$.

(ii) Show that $\bigcup \{ X_k \mid k \in \mathbb{N}_0 \}$ is a sublattice of L.

(iii) Show that $[X] = \bigcup \{ X_k \mid k \in \mathbb{N}_0 \}$.

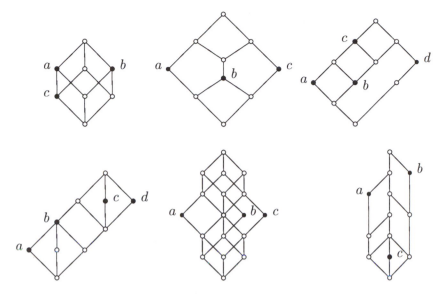

Figure 5.11

(iv) For each of the lattices in Figure 5.11 take X to be the set of shaded elements. Find $[X]$ in each case. As you proceed, label each element in terms of $\vee, \wedge$ and the generators $a, b, c, \ldots .$

5.6 Show that every ideal J of a lattice L is a sublattice. (See Exercise 2.11.)

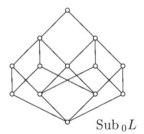

$\mathrm{Sub}\,_0 L$

Figure 5.12

5.7 (i) Draw a labelled diagram of the lattice $\langle \mathrm{Sub}_0 \mathbf{3}; \subseteq \rangle$.

(ii) (a) Find a lattice L such that $\langle \mathrm{Sub}_0 L; \subseteq \rangle$ has the diagram given in Figure 5.12. (Justify your answer by labelling the elements of your guess for L and then drawing a labelled diagram of $\mathrm{Sub}_0 L$.)

(b) Prove that, up to isomorphism, there is only one such lattice L.

5.8 (i) Up to isomorphism there are exactly 10 lattices L for which $1 \leqslant |L| \leqslant 5$. Draw a diagram for each of these lattices.

(ii) Establish the number of sublattices of an n-element chain.

(iii) Prove that, up to isomorphism, there is *at most* one lattice L for which $\langle \mathrm{Sub}_0 L; \subseteq \rangle$ has the diagram given in Figure 5.13.

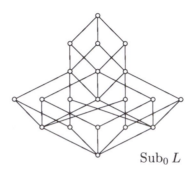
$\mathrm{Sub}_0 L$

Figure 5.13

5.9 Consider again the maps defined in Exercise 1.5. Which of these maps are (i) join-preserving, (ii) homomorphisms?

5.10 Draw the product of the lattices **3** and $\mathbf{2}^2 \oplus \mathbf{1}$ and shade in elements which form a sublattice isomorphic to $\mathbf{1} \oplus (\mathbf{2} \times \mathbf{3}) \oplus \mathbf{1}$.

5.11 Let $f : L \to K$ be a lattice homomorphism.

(i) Show that if $M \in \mathrm{Sub}\, L$ then $f(M) \in \mathrm{Sub}\, K$.

(ii) Show that if $N \in \mathrm{Sub}\, K$ then $f^{-1}(N) \in \mathrm{Sub}\, L$.

5.12 Let L and K be lattices with 0 and 1 and let $M = L \times K$. Show that there exist $a, b \in M$ such that

(a) $\downarrow a \cong L$ and $\uparrow b \cong K$,

(b) $a \wedge b = (0,0)$ and $a \vee b = (1,1)$.

Is the lattice given in Figure 5.14 a product of two lattices each with more than one element? (Justify your answer via a careful case-by-case analysis.)

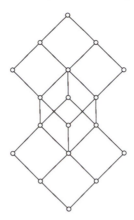

Figure 5.14

5.13 Show that a homomorphism $f\colon L \to K$ is an embedding if and only if $\ker f = \mathbf{0}$.

5.14 Let $\theta \in \operatorname{Con} L$. Show that

$$\theta = \bigvee \{\, \theta(a,b) \mid (a,b) \in \theta \,\}.$$

(Thus every congruence is a join of principal congruences.)

5.15 Let L be a lattice and let $\alpha, \beta \in \operatorname{Con} L$. We say that a sequence $z_0, z_1, \ldots, z_n$ **witnesses** $a(\alpha \vee \beta)b$ if $a = z_0$, $z_n = b$ and $z_{k-1}\alpha z_k$ or $z_{k-1}\beta z_k$ for $1 \leqslant k \leqslant n$. Show that $a(\alpha \vee \beta)b$ if and only if for some $n \in \mathbb{N}$ there exists a sequence $z_0, z_1, \ldots, z_n$ which witnesses $a(\alpha \vee \beta)b$. [Hint. Define a relation θ on L by $a\theta b$ if and only if for some $n \in \mathbb{N}$ there exists a sequence $z_0, z_1, \ldots, z_n$ which witnesses $a(\alpha \vee \beta)b$. Show that (i) $\theta \in \operatorname{Con} L$, (ii) $\alpha \subseteq \theta$ and $\beta \subseteq \theta$, and (iii) if $\alpha \subseteq \gamma$ and $\beta \subseteq \gamma$ for some $\gamma \in \operatorname{Con} L$, then $\theta \subseteq \gamma$.]

5.16 Let L be a lattice and let $a, b, c, d \in L$.

   (i) Show that $\theta(a,b) \subseteq \theta(c,d)$ if and only if $a \equiv b \,(\mathrm{mod}\,\theta(c,d))$.

   (ii) Show that $\theta(a,b) = \theta(a \wedge b, a \vee b)$.

5.17 Consider the lattices L and K in Figure 5.15.

   (i) Show that if $\theta \in \operatorname{Con} L$ and $a \equiv c \,(\mathrm{mod}\,\theta)$, then $b \equiv e \,(\mathrm{mod}\,\theta)$. Find $\theta(a,c)$ and then draw $L/\theta(a,c)$.

   (ii) Show that if $\theta \in \operatorname{Con} K$ and $a \equiv b \,(\mathrm{mod}\,\theta)$, then $b \equiv 0 \,(\mathrm{mod}\,\theta)$. Find $\theta(a,b)$ and then draw $L/\theta(a,b)$.

Figure 5.15

5.18 Find all congruences on $\mathbf{N}_5$, then draw the lattice $\mathrm{Con}\,\mathbf{N}_5$. [Hint. First find all principal congruences on $\mathbf{N}_5$, then show that they are closed under $\vee$ in $\mathrm{Con}\,\mathbf{N}_5$. It then follows by Exercise 5.14 that every congruence on $\mathbf{N}_5$ is principal.]

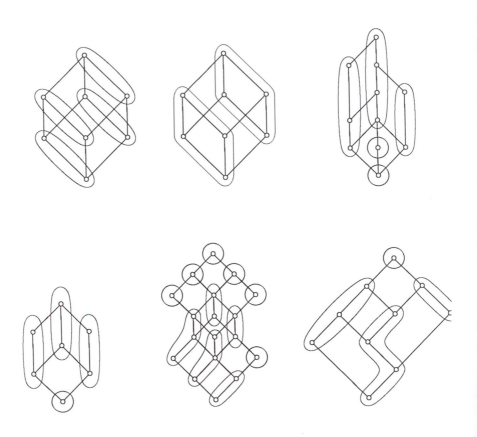

Figure 5.16

5.19 Give an example of a congruence $\theta \in \mathrm{Con}\,\mathbf{4}$ such that θ is not principal.

5.20 A lattice L is called **simple** if it has precisely two congruences, namely $\mathbf{0}$ and $\mathbf{1}$. [Since it can be proved that every lattice can be embedded into a simple lattice, simple lattices can be very complicated!]

   (i) Show that L is simple if and only if for all $a, b \in L$ with $a \neq b$ we have $\theta(a, b) = \mathbf{1}$.

   (ii) Show that each of the following lattices is simple:

   (a) $\mathbf{M_3}$;   (b) $\mathbf{M}_n$ (for $n \geqslant 3$);   (c) $\mathbf{M_{3,3}}$.

   See Figures 2.3, 6.1 and 6.9.

5.21 Draw the diagram of L/θ for each of the congruences shown in Figure 5.16.

5.22 For each of the following lattices L, find all congruences and then draw $\mathrm{Con}\,L$:

   (a) $L = \mathbf{2} \times \mathbf{3}$;   (b) $L = \mathbf{2}^2 \oplus \mathbf{1}$;   (c) $L = \mathbf{4}$.

6

Modular and Distributive Lattices

In Chapter 5 we began an exploration of the algebraic theory of lattices, armed with just enough axioms on $\vee$ and $\wedge$ to ensure that each lattice $\langle L; \vee, \wedge \rangle$ arose from a lattice $\langle L; \leqslant \rangle$ and vice versa. Now we introduce additional identities linking join and meet which hold in many of our examples of lattices.

Lattices satisfying additional identities

Before formally introducing modular and distributive lattices we prove three lemmas which will put the definitions in 6.4 into perspective. The import of these lemmas is discussed in 6.5.

6.1 Lemma. *Let L be a lattice and let $a, b, c \in L$. Then*

(i) $a \wedge (b \vee c) \geqslant (a \wedge b) \vee (a \wedge c)$, *and dually;*

(ii) $a \geqslant c$ *implies* $a \wedge (b \vee c) \geqslant (a \wedge b) \vee c$, *and dually;*

(iii) $(a \wedge b) \vee (b \wedge c) \vee (c \wedge a) \leqslant (a \vee b) \wedge (b \vee c) \wedge (c \vee a)$.

Proof. We leave (i) and (iii) as exercises. (Alternatively, see Exercise 5.2.) By the Connecting Lemma, (ii) is a special case of (i). ∎

6.2 Lemma. *Let L be a lattice. Then the following are equivalent:*

(i) $(\forall a, b, c \in L) \; a \geqslant c \implies a \wedge (b \vee c) = (a \wedge b) \vee c$;

(ii) $(\forall a, b, c \in L) \; a \geqslant c \implies a \wedge (b \vee c) = (a \wedge b) \vee (a \wedge c)$;

(iii) $(\forall p, q, r \in L) \; p \wedge (q \vee (p \wedge r)) = (p \wedge q) \vee (p \wedge r)$.

Proof. The Connecting Lemma gives the equivalence of (i) and (ii). To prove that (iii) implies (ii), assume that $a \geqslant c$ and apply (iii) with $p = a$, $q = b$ and $r = c$. Conversely, assume (ii) holds and that p, q and r are any elements of L. We may put $a = p$, $b = q$ and $c = p \wedge r$ in (ii), and this gives (iii). ∎

6.3 Lemma. *Let L be a lattice. Then the following are equivalent:*

(D) $(\forall a, b, c \in L) \; a \wedge (b \vee c) = (a \wedge b) \vee (a \wedge c)$;

$(D)^{\partial}$ $(\forall p, q, r \in L) \; p \vee (q \wedge r) = (p \vee q) \wedge (p \vee r)$.

Proof. Assume (D) holds. Then, for $p, q, r \in L$,

$$
\begin{aligned}
(p \vee q) \wedge (p \vee r) &= ((p \vee q) \wedge p) \vee ((p \vee q) \wedge r) && \text{(by (D))} \\
&= p \vee (r \wedge (p \vee q)) && \text{(by } (L2)^{\partial} \text{ \& } (L4)^{\partial}) \\
&= p \vee ((r \wedge p) \vee (r \wedge q)) && \text{(by (D))} \\
&= p \vee (q \wedge r) && \text{(by (L1), } (L2)^{\partial} \text{ \& } (L4)^{\partial})
\end{aligned}
$$

so (D) implies (D)$^\partial$. By duality, (D)$^\partial$ implies (D) too.   ∎

6.4 Definitions.   Let L be a lattice.

(i) L is said to be **distributive** if it satisfies the **distributive law**,

$$(\forall a, b, c \in L)\ \ a \wedge (b \vee c) = (a \wedge b) \vee (a \wedge c).$$

(ii) L is said to be **modular** if it satisfies the **modular law**,

$$(\forall a, b, c \in L)\ \ a \geqslant c \implies a \wedge (b \vee c) = (a \wedge b) \vee c.$$

6.5 Remarks.

(1) Lemma 6.1 shows that any lattice is 'half-way' to being both modular and distributive. To establish distributivity or modularity we only need to check an inequality (see 6.6(5) for an example).

(2) Lemma 6.2 serves two purposes. It shows that any distributive lattice is modular. Also it reveals that the rather mysterious modular law can be reformulated as an identity, a fact we need in 6.7. The modular law may be regarded as licence to rebracket $a \wedge (b \vee c)$ as $(a \wedge b) \vee c$, provided $a \geqslant c$. This observation has no mathematical content, but is useful as an *aide memoire*.

(3) Providentially, distributivity can be defined either by (D) or by (D)$^\partial$ (from 6.3). Thus the apparent asymmetry between join and meet in 6.4(i) is illusory. In other words, L is distributive if and only if L^∂ is. An application of the Duality Principle shows that L is modular if and only if L^∂ is.

(4) The universal quantifiers in 6.2 and 6.3 are essential. It is not true, for example, that if particular elements a, b and c in an arbitrary lattice satisfy $a \wedge (b \vee c) = (a \wedge b) \vee (a \wedge c)$, then they also satisfy $a \vee (b \wedge c) = (a \vee b) \wedge (a \vee c)$.

6.6 Examples.

(1) Any powerset lattice $\wp(X)$ is distributive. More generally, any lattice of sets is distributive. In 10.3 we prove the striking result that *every* distributive lattice is isomorphic to a lattice of sets.

(2) Any chain is distributive (enumerate cases, or (slicker) do Exercise 6.3, or (less elementary) use 6.10).

(3) The lattice $\langle \mathbb{N}_0; \mathrm{lcm}, \gcd \rangle$ is distributive. See Exercise 6.6 for a useful necessary and sufficient condition for distributivity which applies neatly to this example.

(4) Let G be a finite group. Exercise 8.15 shows that $\mathrm{Sub}\, G$ is distributive if G is cyclic. The converse is also true but is very much

harder to prove. It requires a more extensive treatment of subgroup lattices than we have space to include.

(5) Our examples of classes of modular lattices come from algebra.

(i) We noted in 2.8 that the set $\mathcal{N}$–Sub G of normal subgroups of a group G forms a lattice under the operations

$$H \wedge K = H \cap K \qquad \text{and} \qquad H \vee K = HK,$$

with $\subseteq$ as the underlying order. Let $H, K, N \in \mathcal{N}$–Sub G, with $H \supseteq N$. Take $g \in H \wedge (K \vee N)$, so $g \in H$ and $g = kn$, for some $k \in K$ and $n \in N$. Then $k = gn^{-1} \in H$, since $H \supseteq N$ and H is a subgroup. This proves that $g \in (H \wedge K) \vee N$. Hence

$$H \wedge (K \vee N) \leqslant (H \wedge K) \vee N.$$

Since the reverse inequality holds in any lattice (by 6.1) the lattice $\mathcal{N}$–Sub G is modular, for any group G.

(ii) It can be shown in a similar way that the lattice of subspaces of a vector space (see 2.19) is modular.

(6) Consider the lattices $\mathbf{M}_3$ (the **diamond**) and $\mathbf{N}_5$ (the **pentagon**) shown in Figure 6.1. The lattice $\mathbf{M}_3$ arose in 1.10 as $\mathcal{N}$–Sub $\mathbf{V}_4$. Hence, by (5)(i), $\mathbf{M}_3$ is modular. It is, however, not distributive. To see this, note that in the diagram of $\mathbf{M}_3$

$$p \wedge (q \vee r) = p \wedge 1 = p \neq 0 = 0 \vee 0 = (p \wedge q) \vee (p \wedge r).$$

The lattice $\mathbf{N}_5$ is not modular (and so also not distributive): in the diagram we have

$$u \geqslant w \text{ and } u \wedge (v \vee w) = u \wedge 1 = u > w = 0 \vee w = (u \wedge v) \vee w.$$

Figure 6.1

These innocent-looking examples turn out to play a crucial role in the identification of non-modular and non-distributive lattices; see the discussion of the $\mathbf{M}_3$–$\mathbf{N}_5$ Theorem below.

6.7 Sublattices, products and homomorphic images. By 5.7, 5.9 and Exercise 5.12, new lattices can be manufactured by forming sublattices, products and homomorphic images. Modularity and distributivity are preserved by these constructions, as follows.

(i) If L is a modular (distributive) lattice, then every sublattice of L is modular (distributive).

(ii) If L and K are modular (distributive) lattices, then $L \times K$ is modular (distributive).

(iii) If L is modular (distributive) and K is the image of L under a homomorphism, then K is modular (distributive).

Here (i) is immediate and (ii) holds because $\vee$ and $\wedge$ are defined co-ordinatewise on products. For (iii) we use the fact that a join- and meet-preserving map preserves any lattice identity; for the modular case we then invoke (i) $\Leftrightarrow$ (iii) in 6.2.

A particularly useful consequence of the above results deserves to be singled out as a proposition.

6.8 Proposition. *If a lattice is isomorphic to a sublattice of a product of distributive (modular) lattices, then it is distributive (modular).*

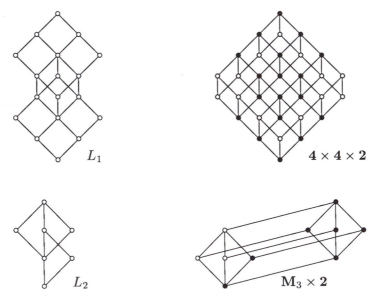

Figure 6.2

6.9 Examples. Consider Figure 6.2. The lattice L_1 is distributive because it is a sublattice of $\mathbf{4} \times \mathbf{4} \times \mathbf{2}$, as indicated. (Any product

of chains is, of course, distributive.) The lattice L_2 is isomorphic to the shaded sublattice of the modular lattice $\mathbf{M}_3 \times \mathbf{2}$ shown alongside and so is itself modular.

The M$_3$–N$_5$ Theorem

We have as yet no way of showing that the distributive law or the modular law is *not* satisfied except a random search for elements for which the law fails. The $\mathbf{M}_3$–$\mathbf{N}_5$ Theorem remedies this in a most satisfactory way. It implies that it is possible to determine whether or not a finite lattice is modular or distributive from its diagram. The first part of the theorem is attributed to R. Dedekind and the second to G. Birkhoff.

We adopt a head-on approach to the proof. This has the disadvantage that it does not reveal why the theorem works. For a more illuminating treatment, beyond the scope of this book, see [24].

We write $M \rightarrowtail L$ to indicate that the lattice L has a sublattice isomorphic to the lattice M. By 5.11(ii), $M \rightarrowtail L$ implies $M \hookrightarrow L$ (recall 1.11).

6.10 The M$_3$–N$_5$ Theorem. *Let L be a lattice.*

(i) *L is non-modular if and only if $\mathbf{N}_5 \rightarrowtail L$.*

(ii) *L is non-distributive if and only if $\mathbf{N}_5 \rightarrowtail L$ or $\mathbf{M}_3 \rightarrowtail L$.*

Proof. By 6.6(6) and 6.7, it will be enough to prove that a non-modular lattice has a sublattice isomorphic to $\mathbf{N}_5$ and that a lattice which is modular but not distributive has a sublattice isomorphic to $\mathbf{M}_3$.

(i)   (ii)

Figure 6.3

Assume that L is not modular. Then there exist elements d, e and f such that $d > f$ and $v > u$, where

$$u = (d \wedge e) \vee f \quad \text{and} \quad v = d \wedge (e \vee f).$$

We aim to prove that $e \wedge u = e \wedge v \ (= p$ say$)$ and $e \vee u = e \vee v \ (= q$ say$)$. Then our required sublattice has elements u, v, e, p, q (which are

clearly distinct). The lattice identities and Exercise 2.8 give

$$v \wedge e = (e \wedge (e \vee f)) \wedge d = d \wedge e \text{ and } u \vee e = (e \vee (d \wedge e)) \vee f = e \vee f.$$

Also, by Lemma 6.1(ii),

$$d \wedge e = (d \wedge e) \wedge e \leqslant u \wedge e \leqslant v \wedge e = d \wedge e$$

and, similarly,

$$e \vee f = u \vee e \leqslant v \vee e \leqslant e \vee f \vee e = e \vee f.$$

This proves our claims and so completes the proof of (i).

Now assume that L is modular but not distributive. We build a sublattice isomorphic to $\mathbf{M}_3$. Take d, e and f such that $(d \wedge e) \vee (d \wedge f) < d \wedge (e \vee f)$. Let

$$
\begin{aligned}
p &:= (d \wedge e) \vee (e \wedge f) \vee (f \wedge d), \\
q &:= (d \vee e) \wedge (e \vee f) \wedge (f \vee d), \\
u &:= (d \wedge q) \vee p, \\
v &:= (e \wedge q) \vee p, \\
w &:= (f \wedge q) \vee p.
\end{aligned}
$$

Clearly $u \geqslant p$, $v \geqslant p$ and $w \geqslant p$. Also, by Lemma 6.1(iii), we have $p \leqslant q$. Hence $u \leqslant (d \wedge q) \vee q = q$. Similarly, $v \leqslant q$ and $w \leqslant q$. Our candidate for a copy of $\mathbf{M}_3$ has elements $\{p, q, u, v, w\}$. We need to check that this subset has the correct joins and meets, and that its elements are distinct.

We shall repeatedly appeal to the modular law, viz.

(M) $a \geqslant c$ implies $a \wedge (b \vee c) = (a \wedge b) \vee c$.

For each application of (M) we underline the elements a and c involved. In the calculations which follow we use the commutative and associative laws many times without explicit mention. We have $d \wedge q = d \wedge (e \vee f)$, by (L4)$^{\partial}$. Also

$$
\begin{aligned}
d \wedge p &= \underline{d} \wedge ((e \wedge f) \vee (\underline{(d \wedge e) \vee (d \wedge f)})) \\
&= (d \wedge (e \wedge f)) \vee ((d \wedge e) \vee (d \wedge f)) \\
&= (d \wedge e) \vee (d \wedge f).
\end{aligned}
$$

Thus $p = q$ is impossible. We conclude that $p < q$.

We next prove that $u \wedge v = p$. We have

$$u \wedge v = ((d \wedge q) \vee \underline{p}) \wedge ((e \wedge q) \vee p)$$

$$= ((e \wedge \underline{q}) \vee \underline{p}) \wedge (d \wedge q)) \vee p \qquad \text{(by (M))}$$

$$= ((q \wedge (e \vee p)) \wedge (d \wedge q)) \vee p \qquad \text{(by (M))}$$

$$= ((e \vee p) \wedge (d \wedge q)) \vee p$$

$$= ((d \wedge (e \vee f)) \wedge (e \vee (f \wedge d))) \vee p \qquad \text{(by (L4) \& (L4)}^{\partial})$$

$$= (d \wedge (\underline{(e \vee f)} \wedge (\underline{e} \vee (f \wedge d)))) \vee p$$

$$= (d \wedge (((e \vee f) \wedge (f \wedge d)) \vee e)) \vee p \qquad \text{(by (M))}$$

$$= (\underline{d} \wedge ((\underline{f \wedge d}) \vee e)) \vee p \qquad \text{(since } d \wedge f \leqslant f \leqslant e \vee f)$$

$$= ((d \wedge e) \vee (f \wedge d)) \vee p \qquad \text{(by (M))}$$

$$= p.$$

In exactly the same way, $v \wedge w = p$ and $w \wedge u = p$. Similar calculations yield $u \vee v = v \vee w = w \vee u = q$.

Finally, it is easy to see that if any two of the elements u, v, w, p, q are equal, then $p = q$, which is impossible. ∎

6.11 Applying the $\mathbf{M_3}$–$\mathbf{N_5}$ Theorem. Consider the lattices in Figure 6.4. The lattices L_1 and L_2 have sublattices isomorphic to $\mathbf{N_5}$, explicitly exhibited, and $\mathbf{M_3} \rightarrowtail L_3$. The $\mathbf{M_3}$–$\mathbf{N_5}$ Theorem implies, immediately and conclusively, that L_1 and L_2 are non-modular and that L_3 is non-distributive.

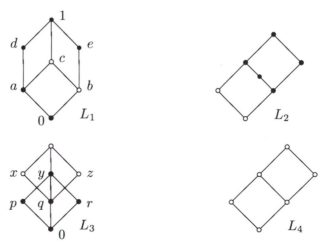

Figure 6.4

It is apparent from the diagrams that $\mathbf{N}_5$ does not embed in L_3 and that neither $\mathbf{N}_5$ nor $\mathbf{M}_3$ embeds in L_4. However, to justify such assertions fully requires a tedious enumeration of cases. For example, suppose $\{u, a, b, c, v\}$, with $u < c < a < v$, $u < b < v$, were a copy of $\mathbf{N}_5$ in L_3. Since L_3 and $\mathbf{N}_5$ both have length 3, we must have $u = 0$ and $v = 1$. By duality and lateral symmetry we may assume without loss of generality that $c = p, a = y$ and $b = r$. But $\{0, p, y, r, 1\}$ is not isomorphic to $\mathbf{N}_5$, $\not{\,\,}$.

To decide whether a given lattice L is non-modular, modular but non-distributive, or distributive, we therefore proceed as follows. If a sublattice of L isomorphic to $\mathbf{N}_5$ ($\mathbf{M}_3$) can be exhibited, then L is non-modular (non-distributive), by the $\mathbf{M}_3$–$\mathbf{N}_5$ Theorem. If a search for a copy of $\mathbf{N}_5$ (of either $\mathbf{N}_5$ or $\mathbf{M}_3$) fails, we conjecture that L is modular (distributive). To substantiate this claim we first try to apply the sublattice-of-a-product technique. Our remarks in Example 6.12 show that there are cases where this is doomed to fail. A fall-back method, relying on a rather tricky proof, is given in Exercise 6.14.

It should be emphasized that the statement of the $\mathbf{M}_3$–$\mathbf{N}_5$ Theorem refers to the occurrence of the pentagon or diamond as a *sublattice* of L; this means that the joins and meets in a candidate copy of $\mathbf{N}_5$ or $\mathbf{M}_3$ must be the same as those in L. In Figure 6.4, the pentagon $K = \{0, a, b, d, 1\}$ in L_1 is *not* a sublattice; $a \vee b = c \notin K$. In the other direction, in applying 6.8 one must be sure to embed the given lattice as a *sublattice*. Thus it would be erroneous to conclude from Figure 5.2(iv) that $\mathbf{N}_5$ is distributive: $\mathbf{N}_5$ sits inside the distributive lattice $\mathbf{2}^3$, but not as a sublattice.

Figure 6.5

6.12 Example. Consider the lattice $\mathbf{M}_{3,3}$ in Figure 6.5. It can be shown (see Exercise 6.22) that if $\mathbf{M}_{3,3} \leqslant A \times B$ for some lattices A and B, then $\mathbf{M}_{3,3} \rightarrowtail A$ or $\mathbf{M}_{3,3} \rightarrowtail B$. Consequently the sublattice-of-a-product technique cannot be applied to $\mathbf{M}_{3,3}$. Nevertheless, $\mathbf{M}_{3,3}$ is modular. To see this, note that for $u \in \{x, y, z\}$, the sublattice $\mathbf{M}_{3,3} \setminus \{u\}$ is isomorphic to L (Figure 6.5) or to its dual, both of which are modular (see 6.9 and 6.5(3)). Thus any sublattice of $\mathbf{M}_{3,3}$ isomorphic to $\mathbf{N}_5$ would need to contain each of x, y and z. This cannot happen, since these points form an antichain.

Exercises

Exercises from the text. Prove Lemma 6.1(i), (iii). Prove that L^∂ is modular if L is (see 6.5(3)).

6.1 Which of the lattices of Figure 6.6 are distributive and which are modular? (Use the $\mathbf{M}_3$–$\mathbf{N}_5$ Theorem and 6.8, as explained in 6.11, to justify your answers.)

6.2 Repeat the previous exercise for the lattices in Figure 5.11.

6.3 (i) Find a set X of least cardinality such that the chain **3** is isomorphic to a sublattice of $\wp(X)$. Conclude that the chains **1, 2** and **3** are distributive.

(ii) Let C be any chain and let $x, y, z \in C$. Show that $x \wedge (y \vee z)$ and $(x \wedge y) \vee (x \wedge z)$ both belong to $\{x, y, z\}$.

(iii) By combining (i) and (ii) show that every chain is distributive.

6.4 Show that $\mathbf{M}_3$ and $\mathbf{N}_5$ are the only non-distributive lattices with fewer than 6 elements.

6.5 Use the $\mathbf{M}_3$–$\mathbf{N}_5$ Theorem to show that every lattice L of length 2 is modular (Exercise 2.20 sought a description of all such lattices). (Hence, in particular, $\mathbf{M}_n$ is modular for all n.)

6.6 (i) Let L be a distributive lattice and let $a, b, c, \in L$. Prove that

$$(a \vee b = c \vee b \ \& \ a \wedge b = c \wedge b) \Longrightarrow a = c. \qquad (*)$$

(ii) Find elements a, b, c in $\mathbf{M}_3$ violating $(*)$. Do the same for $\mathbf{N}_5$.

(iii) Deduce that a lattice L is distributive if and only if $(*)$ holds for all $a, b, c \in L$.

6.7 Guess and then prove a characterization of modularity similar to the characterization of distributivity given in Exercise 6.6.

6.8 (For those with some group theory behind them.)
(i) Prove that no group G satisfies $\mathrm{Sub}\, G \cong \mathbf{N}_5$.

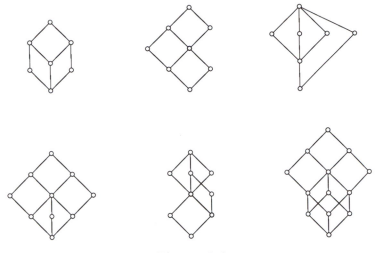

Figure 6.6

(ii) Prove that if a group G satisfies $\mathrm{Sub}\,G \cong \mathbf{M}_3$, then $G \cong \mathbf{V}_4 \cong \mathbb{Z}_2 \times \mathbb{Z}_2$.

6.9   It was seen in Example 2.7(4) and Exercise 2.21 that when $\mathbb{N}_0$ is ordered by division the resulting ordered set is a complete lattice in which finite joins are given by lcm and finite meets by gcd.

(i) Show that $\langle \mathbb{N}_0; \mathrm{lcm}, \mathrm{gcd} \rangle$ is distributive. [Use Exercise 6.6.]

(ii) Let $A = \{3, 5, 7, \dots\}$ be the set of odd primes. Calculate $\bigvee A$ in $\langle \mathbb{N}_0; \preccurlyeq \rangle$. Hence show that $\langle \mathbb{N}_0; \preccurlyeq \rangle$ fails the **Join-Infinite Distributive law**

$$x \wedge \bigvee \{\, y_i \mid i \in I \,\} = \bigvee \{\, x \wedge y_i \mid i \in I \,\}.$$

(Compare this with Exercise 6.10.)

6.10   Let L be a bounded distributive lattice which satisfies (DCC) (and so is complete, by 2.28). Show that L satisfies the **Meet-Infinite Distributive law**

$$x \vee \bigwedge \{\, y_i \mid i \in I \,\} = \bigwedge \{\, x \vee y_i \mid i \in I \,\}.$$

6.11   Use the $\mathbf{M}_3$–$\mathbf{N}_5$ Theorem to show that if L is a distributive lattice, then L^∂ is also distributive. (This yields a non-computational proof of 6.3.)

6.12   (i) Prove that a lattice L is distributive if and only if for each $a \in L$, the map $f_a \colon L \to {\downarrow}a \times {\uparrow}a$ defined by

$$f_a(x) = (x \wedge a, x \vee a) \quad \text{for all } x \in L$$

is a one-to-one homomorphism. [Use Exercise 6.6.]

(ii) Prove that, if L is distributive and possesses 0 and 1, then f_a (as defined in (i)) is an isomorphism if and only if a has a complement (that is, if and only if there exists b such that $a \wedge b = 0$ and $a \vee b = 1$).

6.13   (i) Let L and K be lattices and assume that $\mathbf{N}_5 \rightarrowtail L \times K$. Show that $L \rightarrowtail \mathbf{N}_5$ or $\mathbf{N}_5 \rightarrowtail K$.

(ii) Under the additional assumption that both L and K are modular, repeat (i) with $\mathbf{M}_3$ in place of $\mathbf{N}_5$.

6.14 Assume that L is a lattice, J is an ideal of L, and F is a filter of L (see 9.1 for the definition) and such that $L = J \cup F$ and $J \cap F \neq \varnothing$.

(i) Show that, if L has a sublattice M with $M \cong \mathbf{M}_3$, then either $M \subseteq J$ or $M \subseteq F$. (This does not require $J \cap F \neq \varnothing$.)

(ii) Show that if $x \in J$, $y \in F$ and $x \leqslant y$, then there exists $z \in J \cap F$ such that $x \leqslant z \leqslant y$. [Hint. Consider $x \vee (y \wedge w)$, where $w \in J \cap F$.]

(iii) (a) Prove that if $\mathbf{N}_5 \rightarrowtail L$ then $\mathbf{N}_5 \rightarrowtail J$ or $\mathbf{N}_5 \rightarrowtail F$.

[Hint. Let $N = \{u, a, b, c, v\}$ as in Figure 6.7. The non-trivial case occurs when $u \in L \setminus F$ and $v \in L \setminus J$. Show that (up to duality) it suffices to consider the configuration on the left in Figure 6.7. Apply (ii) to obtain z, then add to the diagram the elements $z \wedge b$ and $a \vee (z \wedge b)$. Show that if $c \vee (z \wedge b) = a \vee (z \wedge b)$, then $\mathbf{N}_5 \rightarrowtail J$; otherwise, add $u \vee (z \wedge b)$ to the diagram and show that then $\mathbf{N}_5 \rightarrowtail F$. Remark 6.5 and Exercise 6.15 will come in useful.]

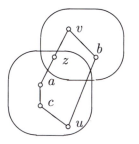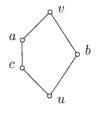

Figure 6.7

(b) Give an example to show that the assumption $J \cap F \neq \emptyset$ is necessary in (a).

(iv) Prove that L is a modular (distributive) lattice if and only if both J and F are modular (distributive) lattices.

(v) Prove that $\mathbf{M}_{3,3}$, as in Example 6.12, and L_2, as in Figure 6.2, are modular.

6.15 (i) Show that $\mathbf{N}_5 \rightarrowtail L$ if and only if there exist five elements $u, a, b, c, v \in L$ such that

    (a) $u < b < a < v$ and $u < c < v$,

    (b) $a \wedge b = u$ and $b \vee c = v$.

(ii) Let $\{u, a, b, c, v\}$ be a sublattice of L isomorphic to $\mathbf{N}_5$ and assume $a', b' \in L$ with $b \leqslant b' < a' \leqslant a$. Show that $\{u, a', b', c, v\}$ is also a sublattice of L isomorphic to $\mathbf{N}_5$.

6.16 Determine the lattices L and K to within isomorphism given that

(a) $L \times K$ contains 20 elements;

(b) L is non-distributive;

(c) K has at least 3 elements;

(d) the greatest element of $L \times K$ covers precisely 4 elements.

6.17 The second half of the proof of the $\mathbf{M}_3$–$\mathbf{N}_5$ Theorem actually shows that, if L is modular and $a, b, c \in L$, then $f \colon \mathbf{M}_3 \to L$ defined by

$$f(0) = p := (a \wedge b) \vee (b \wedge c) \vee (c \wedge a),$$
$$f(1) = q := (a \vee b) \wedge (b \vee c) \wedge (c \vee a),$$
$$f(x) = u := (a \wedge q) \vee p,$$
$$f(y) = v := (b \wedge q) \vee p,$$
$$f(z) = w := (c \wedge q) \vee p,$$

is a homomorphism. Use the fact that $\mathbf{M}_3$ is simple (Exercise 5.20) to show that f is an embedding provided $p \neq q$. (This gives a more sophisticated proof of the last line of the proof of the $\mathbf{M}_3$–$\mathbf{N}_5$ Theorem.)

6.18 Prove that Con L is distributive for every lattice L. [Hint. Let $\alpha, \beta, \gamma \in$ Con L and assume that $a(\alpha \wedge (\beta \vee \gamma))b$. Then $a\alpha b$ and by Exercise 5.15 there is a sequence $a = z_0, z_1, \ldots, z_n = b$ which witnesses $a(\beta \vee \gamma)b$. Define

$$m(x, y, z) := (x \wedge y) \vee (y \wedge z) \vee (z \wedge x),$$

and show that

$$m(x, x, y) = m(x, y, x) = m(y, x, x) = x.$$

Use these identities to show that the sequence

$$a = m(a, b, z_0), m(a, b, z_1), \ldots, m(a, b, z_n) = b$$

witnesses $a((\alpha\wedge\beta)\vee(\alpha\wedge\gamma))b$. Conclude that Con L is distributive.]

6.19 Let J be an ideal of a lattice L and define a relation θ_J on L by

$$\theta_J := \{\, (a, b) \in L^2 \mid (\exists c \in J)\, a \vee c = b \vee c \,\}.$$

Prove that L is distributive if and only if for every ideal J of L, the relation θ_J is a congruence on L and J is a block of the corresponding partition of L.

6.20 Let L be a lattice. Show that L is modular if and only if for all $a, b \in L$ the maps $j_b : x \mapsto x \vee b$ and $m_a : y \mapsto y \wedge a$ are mutually inverse lattice isomorphisms between $[a \wedge b, a]$ and $[b, a \vee b]$. Deduce that, if L is finite, then L is modular if and only if $[a \wedge b, a] \cong [b, a \vee b]$ for all $a, b \in L$. (Here $[c, d] := \{\, x \in L \mid c \leqslant x \leqslant d \,\}$.)

6.21 Show that the lattice Sub $\mathbb{Z}$ of subgroups of the infinite cyclic group $\langle \mathbb{Z}; + \rangle$ is isomorphic to $\langle \mathbb{N}_0; \preccurlyeq \rangle^{\partial}$ and hence is distributive.

6.22 (i) Let L, K_1 and K_2 be lattices and assume that $\varphi \colon L \to K_1 \times K_2$ is an embedding. Define congruences $\theta_1, \theta_2 \in \text{Con } L$ by $\theta_i = \ker(\pi_i \circ \varphi)$ where $\pi_i : K_1 \times K_2 \to K_i$ is the natural projection $(i = 1, 2)$. Show that $\theta_1 \wedge \theta_2 = \mathbf{0}$ in Con L.

(ii) Let $\theta_1, \theta_2 \in \text{Con } L$ with $\theta_1 \wedge \theta_2 = \mathbf{0}$. Show that the map $\psi : L \to L/\theta_1 \times L/\theta_2$ defined by $\psi(a) := ([a]_{\theta_1}, [a]_{\theta_2})$ is an embedding.

(iii) Use (i) to show that each of the following lattices L has the property that if $L \rightarrowtail K_1 \times K_2$, then $L \rightarrowtail K_1$ or $L \rightarrowtail K_2$.

(a) $L = \mathbf{M}_3$; (b) $L = \mathbf{N}_5$; (c) $L = \mathbf{M}_{3,3}$ (see Figure 6.5).

[Hint. If you haven't already done so, you'll first need to calculate Con L in each case (Exercise 5.22). Then use Exercise 5.13.]

7

Boolean Algebras and their applications

A Boolean algebra is a distributive lattice with additional structure which mimics the complementation in a powerset. In this chapter we begin the study of Boolean algebras *qua* lattices. The representation theory of Chapters 8 and 10 carries this further. We also explore the interface between lattice theory and propositional calculus, for the benefit of those who know some logic. Historically, Boolean algebras are inextricably linked to logic, and it is in this context that students in a variety of disciplines encounter them, at levels ranging from primary school to graduate. Many Boolean algebra applications, circuit design for example, are specialist topics which rely on the laws of Boolean algebra but quickly leave lattice theory behind. We only hint at such applications.

Boolean algebras

Our first task is to define complements in an arbitrary lattice.

7.1 Complements. Let L be a lattice with 0 and 1. For $a \in L$, we say $b \in L$ is a **complement** of a if $a \wedge b = 0$ and $a \vee b = 1$. If a has a *unique* complement, we denote this complement by a'.

Assume L is distributive and suppose that b_1 and b_2 are both complements of a. Then

$$b_1 = b_1 \wedge 1 = b_1 \wedge (a \vee b_2) = (b_1 \wedge a) \vee (b_1 \wedge b_2) = (b_1 \wedge b_2).$$

Hence $b_1 \leqslant b_2$ by the Connecting Lemma. Interchanging b_1 and b_2 gives $b_2 \leqslant b_1$. Therefore in a distributive lattice an element can have at most one complement. It is easy to find examples of non-unique complements in non-distributive lattices: look at $\mathbf{M}_3$ or $\mathbf{N}_5$ (Figure 6.1).

A lattice element may have no complement. The only complemented elements in a bounded chain are 0 and 1. In a lattice $\mathfrak{L}$ of subsets of a set X, the element A has a complement if and only if $X \setminus A$ belongs to $\mathfrak{L}$. Thus the complemented elements of $\mathcal{O}(P)$ are the sets which are simultaneously down-sets and up-sets.

7.2 Definition. A lattice L is called a **Boolean lattice** if

  (i) L is distributive,

 (ii) L has 0 and 1,

(iii) each $a \in L$ has a (necessarily unique) complement $a' \in L$.

The following lemma collects together properties of the complement in a Boolean lattice.

7.3 Lemma. *Let L be a Boolean lattice. Then*
(i) $0' = 1$ *and* $1' = 0$;
(ii) $a'' = a$ *for all* $a \in L$;
(iii) *de Morgan's laws: for all* $a, b \in L$,

$$(a \vee b)' = a' \wedge b' \quad \text{and} \quad (a \wedge b)' = a' \vee b';$$

(iv) $a \wedge b = (a' \vee b')'$ *and* $a \vee b = (a' \wedge b')'$ *for all* $a, b \in B$;
(v) $a \wedge b' = 0$ *if and only if* $a \leqslant b$ *for all* $a, b \in L$.

Proof. To prove $p = q'$ in L it is sufficient to prove that $p \vee q = 1$ and $p \wedge q = 0$, since the complement of q is unique. This observation makes the verification of (i)–(iii) entirely routine. Part (iv) follows from (ii) and (iii), while (v) is an easy exercise. ∎

7.4 Boolean algebras. A Boolean lattice was defined to be a special kind of distributive lattice. In such a lattice it is often more natural to regard the distinguished elements 0 and 1 and the unary operation $'$ as an integral part of the structure, with their properties embodied in axioms. Accordingly, a **Boolean algebra** is defined to be a structure $\langle B; \vee, \wedge, ', 0, 1 \rangle$ such that
(i) $\langle B; \vee, \wedge \rangle$ is a distributive lattice,
(ii) $a \vee 0 = a$ and $a \wedge 1 = a$ for all $a \in B$,
(iii) $a \vee a' = 1$ and $a \wedge a' = 0$ for all $a \in B$.

This viewpoint is extended to other concepts from Chapter 5. We say that a subset A of a Boolean algebra B is a **subalgebra** if A is a sublattice of B which contains 0 and 1 and is such that $a \in A$ implies $a' \in A$. Given Boolean algebras B and C, a map $f \colon B \to C$ is a **Boolean homomorphism** if f is a lattice homomorphism which also preserves $0, 1$ and $'$ (that is, $f(0) = 0$, $f(1) = 1$ and $f(a') = (f(a))'$ for all $a \in B$). Lemma 7.5 shows that these conditions are not independent.

7.5 Lemma. *Let $f \colon B \to C$, where B and C are Boolean algebras.*
(i) *Assume f is a lattice homomorphism. Then the following are equivalent:*
  (a) $f(0) = 0$ *and* $f(1) = 1$;
  (b) $f(a') = (f(a))'$ *for all* $a \in B$.
(ii) *If f preserves $'$, then f preserves $\vee$ if and only if f preserves $\wedge$.*

Proof. (i) To confirm that (a) implies (b) use the equations

$$0 = f(0) = f(a \wedge a') = f(a) \wedge f(a'),$$
$$1 = f(1) = f(a \vee a') = f(a) \vee f(a').$$

Conversely, if (b) holds, we have

$$f(0) = f(a \wedge a') = f(a) \wedge (f(a)') = 0,$$
$$f(1) = f(a \vee a') = f(a) \vee (f(a)') = 1.$$

(ii) Assume f preserves $'$ and $\vee$. By Lemma 7.3(iv),

$$f(a \wedge b) = f((a' \vee b')') = (f(a' \vee b'))' = (f(a') \vee f(b'))'$$
$$= ((f(a))' \vee (f(b))')' = f(a) \wedge f(b),$$

for all $a, b \in B$. The converse is proved dually. ∎

Theorem 7.6 is the Boolean algebra version of the Fundamental Homomorphism Theorem; its proof is a straightforward verification.

7.6 Theorem. *Let B and C be Boolean algebras, let f be a Boolean homomorphism of B onto C and define $\theta = \ker f$. Then $\ker f$ is a Boolean congruence (that is, θ is a lattice congruence such that $a \equiv b \,(\mathrm{mod}\,\theta)$ implies $a' \equiv b' \,(\mathrm{mod}\,\theta)$) and the map $g \colon B/\theta \to C$, given by $g([a]_\theta) = f(a)$ for all $[a]_\theta \in B/\theta$, is well defined. Moreover g is an isomorphism between B/θ and C.*

7.7 Examples.

(1) For any set X, let $A' := X \smallsetminus A$ for all $A \subseteq X$. Then the structure $\langle \wp(X); \cup, \cap, ', \varnothing, X \rangle$ is a Boolean algebra known as the **powerset algebra** on X. More generally, an algebra of sets is a Boolean algebra under the set-theoretic operations; by an **algebra of sets** (also known as a **field of sets**) we mean a subalgebra of some powerset algebra $\wp(X)$.

(2) The **finite-cofinite algebra** of the set X is defined to be

$$\mathrm{FC}(X) := \{\, A \subseteq X \mid A \text{ is finite or } X \smallsetminus A \text{ is finite}\,\}.$$

We prove in Chapter 8 that every finite Boolean algebra is isomorphic to some powerset algebra. Finiteness is essential here. The algebra $\mathrm{FC}(\mathbb{N})$ is an example of an infinite Boolean algebra which is not isomorphic to $\wp(X)$ for any X. One way to arrive at this is to consider cardinalities. It is a standard exercise on countable sets to prove that $\mathrm{FC}(\mathbb{N})$ is countable. On the other hand, Cantor's Theorem implies that any powerset is either finite or uncountable. A more lattice-theoretic proof involves completeness. Exercise 2.7 shows that $\mathrm{FC}(\mathbb{N})$ is not complete. But $\wp(X)$ is always complete and an isomorphism must preserve completeness, by Exercise 2.10.

(3) The family of all clopen subsets of a topological space $(X; \mathcal{T})$ is an algebra of sets. Clearly this example will not be of much interest unless X has plenty of clopen sets. The significance of Boolean algebras of this sort emerges in Chapter 10.

(4) For $n \geqslant 1$ the lattice $\mathbf{2}^n$ is lattice-isomorphic to $\wp(\{1, 2, \ldots, n\})$, which is a Boolean algebra. Hence $\mathbf{2}^n$ is a Boolean algebra, with

$$0 = (0, 0, \ldots, 0) \quad \text{and} \quad 1 = (1, 1, \ldots, 1),$$
$$(\varepsilon_1, \ldots, \varepsilon_n)' = (\eta_1, \ldots, \eta_n), \text{ where } \eta_i = 0 \iff \varepsilon_i = 1.$$

The simplest non-trivial Boolean algebra of all is $\mathbf{2} = \{0, 1\}$. It arises frequently in logic and computer science as an algebra of truth values. In such contexts the symbols F and T, or alternatively $\perp$ and $\top$, are used in place of 0 and 1. We have

$$\mathsf{F} \vee \mathsf{F} = \mathsf{F} \wedge \mathsf{F} = \mathsf{F} \wedge \mathsf{T} = (\mathsf{T})' = \mathsf{F},$$
$$\mathsf{T} \wedge \mathsf{T} = \mathsf{F} \vee \mathsf{T} = \mathsf{T} \vee \mathsf{T} = (\mathsf{F})' = \mathsf{T}.$$

In the next section we begin to explore the way Boolean algebras model deductive reasoning involving statements which are assigned value T (true) or F (false).

Boolean terms and disjunctive normal form

Boolean algebras are named after George Boole. In *The laws of human thought*, published in 1840, Boole undertook "to investigate the fundamental laws of those operations of the mind by which reasoning is performed, to give expression to them in the symbolic language of a Calculus and upon this foundation to establish the science of logic and to construct its method ... ". The laws Boole laid down for his Calculus were essentially the laws for a Boolean algebra and he initiated the study of the algebra of propositions, which we discuss next.

7.8 Truth tables; the algebra of propositions. In propositional (or statement) calculus, propositions are designated by **propositional variables**, which take values in $\{\mathsf{F}, \mathsf{T}\}$. Admissible compound statements are formed using **logical connectives**. Connectives include 'and', 'or' and 'not', denoted respectively by our old friends $\wedge, \vee$ and $'$. Another natural connective is 'implies' ($\rightarrow$). Compound statements built from these are assigned the expected truth values according to the truth values of their constituent parts. Thus, for example, $p \wedge q$ has value T if and only if both p and q have value T and $p \rightarrow q$ has value T unless p has value T and q has value F. (Any puzzlement resulting from the fact that $p \rightarrow q$ is true whenever p is false should be dispelled by reading $p \rightarrow q$ as 'if p, then q'.)

Formally, we take an infinite set of propositional variables, denoted $p, q, r, \ldots$, and define a **wff** (or **well-formed formula**) by the rules:

(i) any propositional variable standing alone is a wff (optionally, constant symbols **T** and **F** may also be included as wffs);

(ii) if φ and ψ are wffs, so are $(\varphi \wedge \psi)$, $(\varphi \vee \psi)$, φ', $(\varphi \rightarrow \psi)$ (this clause is suitably adapted if a different set of connectives is used);

(iii) any wff arises from a finite number of applications of (i) and (ii).

Thus $((p \wedge q') \vee r)'$ and $((p' \rightarrow q) \rightarrow ((p' \rightarrow q') \rightarrow p))$ are wffs while $(((p \vee q) \wedge p)$ (invalid bracketing) and $\vee \rightarrow q$ (arrant nonsense) are not. The parentheses may appear to clutter up wffs unnecessarily. Included as dictated by the definition they guarantee non-ambiguity. In practice we drop parentheses where no ambiguity would result, just as if we were writing a string of joins, meets and complements in a lattice.

A wff φ involving the propositional variables $p_1, \ldots, p_n$ defines a **truth function** F_φ of n variables. For a given assignment of values in $\{\mathbf{F}, \mathbf{T}\}$ to $p_1, \ldots, p_n$, simply substitute these values into φ and compute the resulting expression in the Boolean algebra $\{\mathbf{F}, \mathbf{T}\}$ to obtain the value of F_φ. Conventionally, truth functions are presented via **truth tables**, as illustrated in Table 7.1.

p	q	$p \rightarrow q$
T	T	T
T	F	F
F	T	T
F	F	T

p_1	p_2	p_3	$(p_1 \vee p_2)$	$(p_1' \vee p_3)$	$((p_1 \vee p_2) \wedge (p_1' \vee p_3))'$
T	T	T	T	T	F
T	T	F	T	F	T
T	F	T	T	T	F
T	F	F	T	F	T
F	T	T	T	T	F
F	T	F	T	T	F
F	F	T	F	T	T
F	F	F	F	T	T

Table 7.1

Two wffs φ and ψ are called **logically equivalent** (written $\varphi \equiv \psi$) if they define the same truth function, that is, they give rise to the same truth table For the purposes of deductive reasoning, logically equivalent wffs are essentially the same. It is an easy exercise on truth tables to prove that, for any wffs φ and ψ,

$$(\varphi \wedge \psi) \equiv (\varphi' \vee \psi')', \quad (\varphi \vee \psi) \equiv (\varphi' \wedge \psi')',$$
$$(\varphi \to \psi) \equiv (\varphi' \vee \psi), \quad (\varphi \wedge \psi) \equiv (\varphi \to \psi')'.$$

A proof by induction on the number of connectives then shows that any wff built using $\vee, \wedge$ and $'$ is logically equivalent to one built using $\to$ and $'$, and vice versa. Therefore, up to logical equivalence, we arrive at the same set of wffs whether we take $\{\vee, \wedge, ', \to\}$, just $\{\to, '\}$ or just $\{\vee, \wedge, '\}$ as the basic set of connectives. The choice of $\{\to, '\}$ is the most natural for studying logic, while $\{\vee, \wedge, '\}$ brings out the connections with Boolean algebras.

The set of wffs, with $\vee, \wedge$ and $'$ as operations, closely resembles a Boolean lattice. The axioms do not hold if $=$ means 'is the same wff as', but it is a routine matter to show that they all hold if $=$ is read as 'is logically equivalent to'. For example, to establish (L4) note that $\varphi \vee (\varphi \wedge \psi)$ takes value **T** if and only if φ does, so $\varphi \vee (\varphi \wedge \psi) \equiv \varphi$. If **F** and **T** are included as wffs, to serve as 0 and 1, we obtain a Boolean algebra. We meet this **algebra of propositions** again, in a more formal guise, in the next section.

7.9 Boolean terms. We used the Boolean symbols, representing logical connectives, to build the formulae of propositional calculus from propositional variables. This construction, freed from the trappings of logic, has a variety of applications. We define the class **BT** of **Boolean terms** (or **Boolean polynomials**) as follows. Let S be a set of variables, whose members will be denoted by letters such as $x, y, z, x_1, x_2, \ldots$, and let $\vee, \wedge, ', 0, 1$ be the symbols used to axiomatize Boolean algebras. Then

(i) $0, 1 \in$ **BT** and $x \in$ **BT** for all $x \in S$,

(ii) if $p, q \in$ **BT** then $(p \vee q), (p \wedge q)$ and p' belong to **BT**,

(iii) every element of **BT** is an expression formed by a finite number of applications of (i) and (ii).

A Boolean term p whose variables are drawn from among $x_1, \ldots, x_n$ will be written $p(x_1, \ldots, x_n)$. Examples of Boolean terms, illustrating the building process, are

$$1, \ x, \ y, \ y', \ (x \vee y'), \ (1 \wedge (x \vee y')), \ (1 \wedge (x \vee y'))'.$$

Just as numbers may be substituted into 'ordinary' polynomials, elements of any Boolean algebra B may be substituted for the variables of a Boolean term to yield an element of B. In particular we may take $B = \mathbf{2}$. Thence every Boolean term $p(x_1, \ldots, x_n)$ defines a map $F_p \colon \mathbf{2}^n \to \mathbf{2}$. The map F_p associated with p can be specified by a 'truth table' in just the same way as a wff determines a truth function; the only difference is that each entry of the table is 0 or 1, instead of **F** or **T**.

We say that $p(x_1, \ldots, x_n)$ and $q(x_1, \ldots, x_n)$ are **equivalent**, and write $p \equiv q$, if p and q have the same truth function, that is, $F_p = F_q$. It is easy to see that, for instance, $(x \wedge y')' \equiv (x' \vee y)$; just check that both sides give the same truth table. Note that the right hand side can be obtained from the left by applying the laws of Boolean algebra, treating the variables as though they were Boolean algebra elements:

$$(x \wedge y')' = (x' \vee y'') = (x' \vee y).$$

In general, whenever $q(x_1, \ldots, x_n)$ can be obtained from $p(y_1, \ldots, y_n)$ by the laws of Boolean algebra, we have $p \equiv q$. We see in 7.13 that the converse is also true. Where removal of parentheses from a Boolean term would, up to equivalence, not result in ambiguity, we shall omit the parentheses. For example, we shall write $x \vee y \vee z$ in place of either $(x \vee (y \vee z))$ or $((x \vee y) \vee z)$.

We prove in 7.12 that *every* map from $\mathbf{2}^n$ to $\mathbf{2}$ coincides with F_p for some Boolean term, p, in n variables. This is a surprising and important theorem. To motivate both the theorem and its proof, we preface it with a brief discussion of one application.

7.10 Boolean terms and computer architecture. The design of a computer system may be viewed hierarchically: from customer requirement, down in multiple stages through high level programming language, machine code and integrated circuits to semiconductors. The circuits are carried on chips of silicon and consist of interconnected groups of transistors. A crucial feature of transistors is that although subjected to continuously varying voltages, they either allow current to pass to the best of their ability or not at all, and so act as electrical switches. Transistors are linked to create **gates**. A gate recognizes only two levels of voltage: high (denoted 1) and low (denoted 0). It may be regarded as having n inputs each taking value 0 or 1, and having one or more outputs, each taking value 0 or 1 (depending on the combination of inputs). A basic kit for constructing circuits consists of

AND gate two inputs; output 1 if and only if both inputs are 1,
OR gate two inputs; output 1 if and only if either input is 1,
NOT gate one input; output 0 if and only if the input is 1.

Figure 7.1(i) shows a stylized representation of these gates and 7.1(ii) a **gate diagram** for a circuit. The same input-output behaviour results from ordinary electric switches wired in series (for an AND gate) and in parallel (for an OR gate). Thus gate diagrams are really what in olden, pre-transistor, days were known as **series-parallel switching circuits**.

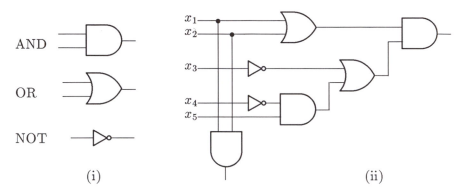

AND

OR

NOT

(i)   (ii)

Figure 7.1

Clearly AND, OR and NOT gates mimic $\wedge$, $\vee$ and $'$ acting on $\{0, 1\}$. Thus a gate diagram with k outputs corresponds to a k-element set of Boolean terms. The 2–output diagram in Figure 7.1(ii) corresponds to the 2-element set of terms $\{x_1 \wedge x_2, (x_1 \vee x_2) \wedge (x_3' \vee (x_4' \wedge x_5))\}$.

The problem of constructing a gate diagram to model a circuit with specified characteristics is just that of finding a Boolean term with a given truth function. Theorem 7.12 solves this problem, but in a way which is in general highly redundant. In circuit design this may be very undesirable, A complicated Boolean term may lead to a circuit which is costly (many connectives entail many gates), hard to realize compactly on a chip (if the term is 'irregular'), or slow (if the term involves many sub-terms or long strings of joins or meets). This difficulty can in theory be overcome by using the laws of Boolean algebra to replace a complicated Boolean term by a simpler equivalent one; see Example 7.11(2). However a single chip may have a million transistors, and direct implementation of complicated circuits would be impractical and wastefully repetitive. Instead a modular approach is adopted. At the level above gate diagrams in the design hierarchy comes microprogramming. This deals with the implementation of relatively simple modular components (adders, memories, etc.) from which, in turn, more complex processing units are constructed. Effective system design depends on good communication between adjacent levels in the hierarchy. Thus gate diagrams cannot be divorced from microprogramming (above) and the wiring of transistors (below). The use of Boolean terms is enmeshed with the methodology of these related topics, so that we can only illustrate it in a limited way. A fuller account can be found in [48], for example.

7.11 Examples.

(1) Each stage in the binary addition of two numbers involves addition modulo 2, with a 'carry' if two 1s are added. For example, suppose we wish to add 6, represented as 110, to 3, represented as 011. We add the final bits 0 and 1 to give 1 and no carry, then add the penultimate bits to give 0 with a carry 1, and finally add the leftmost bits, remembering to include the carry. This gives the expected result, $1001 = 9$. A circuit (**an adder**) to execute this procedure can be built from components known as **half-adders**, where each half-adder carries out a single 'sum and carry' operation on a pair of bits. Input-output for a half-adder is shown in Table 7.2.

x	y	sum	$carry$
0	0	0	0
0	1	1	0
1	0	1	0
1	1	0	1

Table 7.2

It is immediate that *carry* is given by the Boolean term $x \wedge y$. For *sum* we require a term $p(x, y)$ which takes value 1 when exactly one of x, y takes value 1. Looking at lines 2 and 3 of the table we see that $x \wedge y'$ and $x' \wedge y$ have this property; no other meets of pairs do. It is then routine to verify that the term $(x \wedge y') \vee (x' \wedge y)$ has exactly the truth table we require for *sum*. The associated gate diagram is shown in Figure 7.2. This can be simplified by having tailor-made gates for other 2-variable truth functions; here an XOR gate is wanted, modelling the exclusive form of OR.

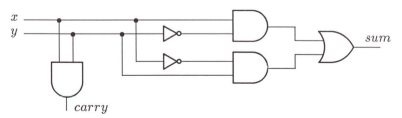

Figure 7.2

(2) Our second example concerns a circuit to execute a logical operation

x	y	z	$p(x, y, z)$
0	0	0	0
0	0	1	1
0	1	0	0
0	1	1	1
1	0	0	0
1	0	1	0
1	1	0	1
1	1	1	1

Table 7.3

as opposed to an arithmetical one. We seek $p(x, y, z)$ having the truth function given in Table 7.3.

Notice that if $x = 1$ then $p(x, y, z)$ must take the same value as y and otherwise must take the value of z. We are therefore modelling `if-then-else`, for which an appropriate term is $(x \wedge y) \vee (x' \wedge z)$. Let us now see how we can arrive at this by the technique we used in the previous example. We want $p(x, y, z) = 1$ for the combinations of truth values in rows 2, 4, 7 and 8. The terms $x' \wedge y' \wedge z$, $x' \wedge y \wedge z$, $x \wedge y \wedge z'$ and $x \wedge y \wedge z$ give value 1 on these rows. Taking the join of these we have a candidate for $p(x, y, z)$, namely

$$(x' \wedge y' \wedge z) \vee (x' \wedge y \wedge z) \vee (x \wedge y \wedge z') \vee (x \wedge y \wedge z)$$
$$\equiv ((x' \wedge z) \wedge (y \vee y')) \vee ((x \wedge y) \wedge (z \wedge z')) \equiv (x \wedge y) \vee (x' \wedge z).$$

The construction in the following proof generalizes that used above. Take the truth table associated with a given truth function $F \colon \mathbf{2}^n \to \mathbf{2}$. For each row (element of $\mathbf{2}^n$) on which F has value 1, form the meet of n symbols by selecting for each variable x either x or x' depending on whether x has value 1 or 0 in that row. The join of these terms, p, is such that $F = F_p$.

7.12 Theorem. *Every map $F : \mathbf{2}^n \to \mathbf{2}$ coincides with F_p for some Boolean term $p(x_1, \ldots, x_n)$. A suitable term p may be described as follows. For $\mathbf{a} = (a_1, \ldots, a_n) \in \mathbf{2}^n$, define $p_{\mathbf{a}}(x_1, \ldots, x_n)$ by*

$$p_{\mathbf{a}}(x_1, \ldots, x_n) = x_1^{\varepsilon_1} \wedge \ldots \wedge x_n^{\varepsilon_n} \text{ where } x_j^{\varepsilon_j} = \begin{cases} x_j & \text{if } a_j = 1, \\ x_j' & \text{if } a_j = 0. \end{cases}$$

Then define

$$p(x_1, \ldots, x_n) = \bigvee \{ p_{\mathbf{a}}(x_1, \ldots, x_n) \mid F(\mathbf{a}) = 1 \}.$$

Proof. Let $\mathbf{a} = (a_1, \ldots, a_n)$ and $\mathbf{b} = (b_1, \ldots, b_n)$. We have carefully chosen the term $p_{\mathbf{a}}(x_1, \ldots, x_n)$ so that

$$p_{\mathbf{a}}(b_1, \ldots, b_n) = \begin{cases} 1 & \text{if } \mathbf{b} = \mathbf{a}, \\ 0 & \text{if } \mathbf{b} \neq \mathbf{a}. \end{cases}$$

We claim that $F_p((b_1, \ldots, b_n)) = F((b_1, \ldots, b_n))$ for every $(b_1, \ldots, b_n) \in \mathbf{2}^n$. Assume $F((b_1, \ldots, b_n)) = 1$. Then

$$
\begin{aligned}
p(b_1, \ldots, b_n) &= \bigvee \{\, p_{\mathbf{a}}(b_1, \ldots, b_n) \mid F(\mathbf{a}) = 1 \,\} \\
&\geqslant p_{\mathbf{b}}(b_1, \ldots, b_n) && (\text{since } F(\mathbf{b}) = 1) \\
&= 1 && (\text{by the above}).
\end{aligned}
$$

Thus $F(\mathbf{b}) = 1$ implies $p(b_1, \ldots, b_n) = 1$. Now assume $F((b_1, \ldots, b_n)) = 0$. Then $F(\mathbf{a}) = 1$ implies $\mathbf{b} \neq \mathbf{a}$, so $p_{\mathbf{a}}(b_1, \ldots, b_n) = 0$. Therefore

$$p(b_1, \ldots, b_n) = \bigvee \{\, p_{\mathbf{a}}(b_1, \ldots, b_n) \mid F(\mathbf{a}) = 1 \,\} = 0. \qquad \blacksquare$$

7.13 Disjunctive normal form. A Boolean term $p(x_1, \ldots, x_n)$ is said to be in **full disjunctive normal form**, or DNF, if it is a join of m meets each of the form $x_1^{\varepsilon_1} \wedge \ldots \wedge x_n^{\varepsilon_n}$. By definition, x^{ε} equals x if $\varepsilon = 1$ and x' if $\varepsilon = 0$; terms of the form x^{ε} are known as **literals**.

Theorem 7.12 implies that any Boolean term is equivalent to a term in DNF (in the setting of propositional calculus this is just the classic result that any wff is logically equivalent to a wff in DNF). Each truth function uniquely determines, and is determined by, a DNF term, so $p \equiv q$ in **BT** if and only if each of p and q is equivalent to the same DNF, in the sense that every meet of literals occurring in the DNF of p occurs in the DNF of q and vice versa. We have already remarked that applying the laws of Boolean algebra to a Boolean term yields an equivalent term. We observe that this process can be used to reduce any term $p(x_1, \ldots, x_n)$ to DNF, as outlined below.

(i) Use de Morgan's laws to reduce $p(x_1, \ldots, x_n)$ to literals combined by joins and meets.

(ii) Use the distributive laws repeatedly, with the lattice identities, to obtain a join of meets of literals.

(iii) Finally we require each x_i to occur, either primed or not, once and once only in each meet term. This is achieved by dropping any terms containing both x_i and x_i', for any i. If neither x_j nor x_j' occurs in $\bigwedge_{k \in K} x_k^{\varepsilon_k}$, note that

$$\bigwedge_{k \in K} x_k^{\varepsilon_k} \equiv \left(\bigwedge_{k \in K} x_k^{\varepsilon_k} \right) \wedge (x_j \vee x_j') \equiv \left(\bigwedge_{k \in K} x_k^{\varepsilon_k} \wedge x_j \right) \vee \left(\bigwedge_{k \in K} x_k^{\varepsilon_k} \wedge x_j' \right).$$

Repeating this for each missing variable we arrive at a term in DNF.

This process shows that a term may be converted to DNF by the laws of Boolean algebra. We may therefore assert that $p \equiv q$ if and only if $p(a_1, \ldots, a_n) = q(a_1, \ldots, a_n)$ for any elements $a_1, \ldots, a_n$ in any Boolean algebra B, and it is possible to test whether this is true by reducing each of p and q to DNF; see Exercise 7.17.

For the term $((p_1 \vee p_2) \wedge (p_1' \vee p_3))'$ the truth table in Table 7.1 and Theorem 7.12 give the DNF

$$(p_1' \wedge p_2' \wedge p_3) \vee (p_1 \wedge p_2 \wedge p_3') \vee (p_1 \wedge p_2' \wedge p_3') \vee (p_1' \wedge p_2' \wedge p_3').$$

For comparison, try obtaining this by the process above.

Finally we give a theorem which is essentially a reformulation of Theorem 7.12.

7.14 Theorem. *Let B be the Boolean algebra 2^{2^n}. Then B is generated by n elements, in the sense that there exists an n-element subset X of B such that the smallest Boolean subalgebra of B containing X is B.*

Proof. Identify B with $\wp(2^n)$, where each element of 2^n is regarded as an n–tuple $\mathbf{a} = (a_1, \ldots, a_n)$ with each $a_i = 0$ or 1. For $i = 1, \ldots, n$, define $e_i := \{ (a_1, \ldots, a_n) \in 2^n \mid a_i = 1 \}$ and let $X := \{e_1, \ldots, e_n\}$. Then for each $\mathbf{a} = (a_1, \ldots, a_n) \in 2^n$ we have

$$\{\mathbf{a}\} = \bigcap_{i=1}^{n} Y_i(\mathbf{a}) \text{ where } Y_i(\mathbf{a}) = \begin{cases} e_i & \text{if } a_i = 1, \\ e_i' & \text{if } a_i = 0. \end{cases}$$

Each non-empty element of B is a union of singletons, $\{\mathbf{a}\}$, and hence expressible as a join of meets of elements of the form e_i or e_i'; noting $\varnothing = e_1 \cap e_1'$ takes care of the empty set. ∎

Meet LINDA: The Lindenbaum algebra

This optional section deals with an important fragment of mathematical logic and the part Boolean algebras play in it. We do not claim to be presenting a primer on formal logic, and those unfamiliar with the subject are referred to standard texts for motivation and background.

There are two quite different approaches to propositional calculus. One is the **semantic** one, based on assignments of truth values, which we discussed in 7.8. In this approach a wff is said to be true if its truth function always takes value **T**. Such a wff is called a **tautology**.

The alternative is a **syntactic** approach, based on a formal deduction system in which a wff is declared to be true if it can be derived from a set of axioms by specified deduction rules. One such system is outlined below.

The reconciliation of these two approaches is discussed in 7.16.

7.15 The formal system L. A **deduction system** consists of

(i) a set of formulae,

(ii) a subset of the formulae designated as axioms,

(iii) a finite set of deduction rules.

The system **L** of propositional calculus is defined as follows.

(i) The formulae are the wffs of propositional calculus, with $\to$ and $\neg$ as connectives. (In this section we use the notation $\neg\varphi$ for the negation of φ, rather than φ', and introduce $(\varphi \vee \psi)$ as shorthand for $(\neg\varphi \to \psi)$ and $(\varphi \wedge \psi)$ as shorthand for $\neg(\varphi \to \neg\psi)$.)

(ii) The axioms of **L** are all wffs of the form

(A1) $(\varphi \to (\psi \to \varphi))$,

(A2) $((\varphi \to (\psi \to \chi)) \to ((\varphi \to \psi) \to (\varphi \to \chi)))$,

(A3) $((\neg\varphi \to \neg\psi) \to (\psi \to \varphi))$,

where φ, ψ and χ are any wffs.

(iii) There is a single deduction rule, **modus ponens**:

(MP) From φ and $(\varphi \to \psi)$ deduce ψ.

A **proof** in **L** is a finite sequence of formulae of which each is either an axiom or is obtained from two previous ones by (MP). A **theorem** of **L** is the last formula in a proof (in other words, it is the culmination of a proof). If φ is a theorem of **L** we write $\vdash_{\mathbf{L}} \varphi$. Theorems *of* **L** must not be confused with theorems *about* the system **L**, which are usually called **metatheorems**.

7.16 Semantics versus syntax. We now have two classes of wffs with a claim to be called true:

(i) the tautologies (semantically true);

(ii) the theorems of the formal system **L** (syntactically true).

We can say that **L** successfully models deductive reasoning with propositions if it is

sound, that is, every theorem is a tautology, and

adequate, that is, every tautology is a theorem.

The major metatheorems of propositional calculus are the Soundness Theorem and the Adequacy Theorem, asserting respectively that **L** is sound and **L** is adequate. Of these, the Soundness Theorem is much the more elementary. It works because (a) each axiom is a tautology (a routine verification) and (b) from tautologies φ and $\varphi \to \psi$, (MP) yields another tautology, ψ. The Adequacy Theorem is far more subtle, and it is only the judicious choice of axioms that makes it work. Indeed, it is remarkable that it does. As anyone who has tried will appreciate, establishing that a particular wff of **L** is a theorem can be a tricky business. The Adequacy Theorem says that a given wff is a theorem of **L** if it is a tautology, and this may be confirmed or refuted by writing down the truth table, a purely mechanical process which can be carried out in a finite number of steps.

In Chapter 9 we derive the Adequacy Theorem as a corollary of a theorem about Boolean algebras. In preparation we need to associate a Boolean algebra with the formal system **L**, which is a syntactic counterpart of the Boolean algebra to which we alluded, rather informally, at the end of 7.8. Before discussing these algebras we recall some facts from propositional calculus which we have not so far needed.

7.17 Valuations. A map v from wffs to $\{\mathbf{F}, \mathbf{T}\}$ is a **valuation** if

(i) $v(\neg\varphi) = \mathbf{T}$ if and only if $v(\varphi) = \mathbf{F}$,

(ii) $v(\varphi \to \psi) = \mathbf{T}$ unless $v(\varphi) = \mathbf{T}$ and $v(\psi) = \mathbf{F}$.

An equivalent definition requires v to preserve $\vee, \wedge$ and $'$ (interpreted on the set of wffs as connectives and on $\{\mathbf{F}, \mathbf{T}\}$ as in 7.7). Elementary lemmas assert that any assignment of truth values to the propositional variables extends in a unique way to a valuation and that a wff φ is a tautology if and only if $v(\varphi) = \mathbf{T}$ for all valuations v.

Given wffs φ and ψ, we say φ **logically implies** ψ, and write $\varphi \models \psi$, if whenever $v(\varphi) = \mathbf{T}$ for a valuation v, then $v(\psi) = \mathbf{T}$. Note that $\varphi \equiv \psi$ if and only if $\varphi \models \psi$ and $\psi \models \varphi$.

7.18 The Lindenbaum algebra, LINDA. Define equivalence relations $\sim_\models$ and $\sim_\vdash$, **semantic equivalence** and **syntactic equivalence**, on wffs by

$$\varphi \sim_\models \psi \text{ if and only if } \varphi \equiv \psi,$$
$$\varphi \sim_\vdash \psi \text{ if and only if } \vdash_{\mathbf{L}} (\varphi \to \psi) \ \& \ \vdash_{\mathbf{L}} (\psi \to \varphi).$$

Given the Soundness and Adequacy Theorems, it is an easy exercise to show that $\sim_\models$ and $\sim_\vdash$ are actually the same relation. However en route to proving the Adequacy Theorem we must treat these relations independently. Let $\sim$ denote either $\sim_{\models}$ or $\sim_{\vdash}$, let $[\varphi]$ be the equivalence class of φ under $\sim$ and denote the set of $\sim$–equivalence classes by LA or, where we need to specify which relation is being used, $\text{LA}_{\models}$ or $\text{LA}_{\vdash}$.

We show that, for either choice of $\sim$, there are natural operations making LA into a Boolean algebra. The most economical route is to define an order relation on LA, to show this makes LA a lattice and finally to show that this lattice is Boolean. All the verifications required are much easier for $\sim_{\models}$ than for $\sim_{\vdash}$. This is only to be expected. In the former case only logical equivalence and implication are involved. In the latter, it is necessary to show that many wffs are theorems of **L**. We only give an indication of the steps, but energetic readers familiar with **L** should be able to complete the proofs. The Deduction Theorem, a standard metatheorem of propositional calculus not relying on the Adequacy Theorem, is an extremely useful tool.

Define $\leqslant$ on $\text{LA}_{\models}$ by

$$[\varphi] \leqslant [\psi] \text{ if and only if } \varphi \models \psi$$

and on $\text{LA}_{\vdash}$ by

$$[\varphi] \leqslant [\psi] \text{ if and only if } \vdash_{\mathbf{L}} (\varphi \to \psi).$$

It can be checked in either case that $\leqslant$ is well defined, that is, $[\varphi] = [\varphi_1]$, $[\psi] = [\psi_1]$ and $[\varphi] \leqslant [\psi]$ together imply $[\varphi_1] \leqslant [\psi_1]$. Further, $\leqslant$ is an order relation. In $\langle \text{LA}; \leqslant \rangle$ there are greatest and least elements,

$$1 = \begin{cases} [\varphi], \text{ where } \varphi \text{ is any tautology} & (\text{for } \sim_{\models}), \\ [\varphi], \text{ where } \varphi \text{ is such that } \vdash_{\mathbf{L}} \varphi & (\text{for } \sim_{\vdash}), \end{cases}$$

and 0, obtained similarly, with $[\varphi]$ replaced by $[\neg\varphi]$.

The next step is to define join, meet and complement on LA. Let

$$[\varphi] \vee [\psi] := [\varphi \vee \psi], \quad [\varphi] \wedge [\psi] := [\varphi \wedge \psi], \quad [\varphi]' := [\neg\varphi].$$

We claim that

(i) $\langle \text{LA}; \leqslant \rangle$ is a lattice with join and meet given by $\vee$ and $\wedge$;

(ii) $\langle \text{LA}; \vee, \wedge \rangle$ is distributive;

(iii) $[\varphi] \vee [\varphi]' = 1$ and $[\varphi] \wedge [\varphi]' = 0$.

Some guidance on checking these claims for $\text{LA}_{\vdash}$ is called for. To show, for example, that $[\varphi \vee \psi]$ is the least upper bound of $[\varphi]$ and $[\psi]$ with respect to $\leqslant$, we need

$$\vdash_{\mathbf{L}} (\varphi \to (\neg\varphi \to \psi)),$$
$$\vdash_{\mathbf{L}} (\psi \to (\neg\varphi \to \psi)),$$
$$\vdash_{\mathbf{L}} ((\varphi \to \chi) \to ((\psi \to \chi) \to ((\neg\varphi \to \psi) \to \chi))) \quad \text{for any wff } \chi.$$

The first of these is a well-known theorem and the second is an instance of an axiom. The third is easily obtained using the Deduction Theorem and the following theorems of **L**:

$$\vdash_{\mathbf{L}} ((\alpha \rightarrow \beta) \rightarrow (\neg\beta \rightarrow \neg\alpha)), \quad \vdash_{\mathbf{L}} ((\neg\alpha \rightarrow \neg\beta) \rightarrow ((\neg\alpha \rightarrow \beta) \rightarrow \alpha)).$$

We conclude that each of $\langle \mathrm{LA}_{\models}; \vee, \wedge, ', 0, 1 \rangle$ and $\langle \mathrm{LA}_{\vdash}; \vee, \wedge, ', 0, 1 \rangle$ is a Boolean algebra. The former algebra was introduced, without the formality of equivalence classes, in 7.8.

Once the Adequacy Theorem is established we know that these are actually the same Boolean algebra. It is known as the **Lindenbaum algebra**, or, to its friends, as LINDA.

7.19 Valuations and homomorphisms. There is a connection between valuations and Boolean homomorphisms from $\mathrm{LA}_{\models}$ or $\mathrm{LA}_{\vdash}$ to **2**. Since valuation is a semantic concept it might seem more natural to consider $\mathrm{LA}_{\models}$. However working instead with $\mathrm{LA}_{\vdash}$ provides the key to the Boolean algebra proof of the Adequacy Theorem in 9.15. Assume that v is a valuation. The Soundness Theorem implies that f_v, given by

$$f_v([\varphi]) = \begin{cases} 1 & \text{if } v(\varphi) = \mathbf{T}, \\ 0 & \text{if } v(\varphi) = \mathbf{F}, \end{cases}$$

is a well-defined map from $\mathrm{LA}_{\vdash}$ to **2**. It is routine to show that f_v is a Boolean homomorphism and that every Boolean homomorphism from $\mathrm{LA}_{\vdash}$ to **2** arises in this way from some valuation.

Exercises

Exercises from the text. Complete the proof of Lemma 7.3 (i)–(iv) (see Exercise 7.7 for (v)). Prove Theorem 7.6.

7.1   Let $n \in \mathbb{N}_0$ and consider the sublattice $L = {\downarrow}n$ of $\langle \mathbb{N}_0; \preccurlyeq \rangle$. When does $m \in L$ have a complement? Give a formula for m' in L when it exists.

Characterize those $n \in \mathbb{N}_0$ such that $L = {\downarrow}n$ is a Boolean lattice.

7.2   For a Boolean lattice B and $a, b \in B$ such that $a \leqslant b$, show that the interval sublattice

$$[a, b] := {\uparrow}a \cap {\downarrow}b = \{\, x \in B \mid a \leqslant x \leqslant b \,\}$$

is a Boolean lattice. When is $[a, b]$ a Boolean subalgebra of B? [Hint. First show that for any distributive lattice L the map $f \colon L \rightarrow L$, given by $f(x) := (x \vee a) \wedge b$, is a homomorphism. Then calculate $f(L)$.]

7.3 Use Exercise 6.12 to give a proof by induction on $|B|$ that $|B| = 2^n$ for every finite Boolean lattice B.

7.4 Let S be a set and $f: S \to S$ any map. A subset A of S is f–**invariant** if $x \in A$ implies $f(x) \in A$; note that $\varnothing$ is f–invariant. Denote the set of all f–invariant subsets of S by $\mathfrak{L}(S, f)$.

(i) Show that $\mathfrak{L}(S, f)$ is a lattice of sets.

(ii) Show that if S is finite and f is bijective, then $\mathfrak{L}(S, f)$ is an algebra of sets. Give an example to show that in general the finiteness of S is necessary here.

(iii) Prove that if $\mathfrak{L}(S, f)$ is an algebra of sets then f is bijective.

7.5 Let B be the family of all finite unions of subintervals of $\mathbb{R}$ of the following types: $(-\infty, a)$, $[a, b)$, and $[b, \infty)$, where $-\infty < a < b < \infty$, together with $\varnothing$. Show that B is a Boolean subalgebra of the powerset algebra $\wp(\mathbb{R})$ and that B has no atoms. (An element a of B is called an **atom** if $0 \prec a$; see 8.1.)

7.6 Show that the following hold in all Boolean algebras:

(i) $(a \wedge b) \vee (a' \wedge b) \vee (a \wedge b') \vee (a' \wedge b') = 1$;

(ii) $(a \wedge b) \vee (a' \wedge c) = (a \wedge b) \vee (a' \wedge c) \vee (b \wedge c)$;

(iii) $a = b \iff (a \wedge b') \vee (a' \wedge b) = 0$;

(iv) $a \wedge b \leqslant c \vee d \iff a \wedge c' \leqslant b' \vee d$.

7.7 Let B be an ordered set such that $a \wedge b$ exists in B for all $a, b \in B$. Show that B is a Boolean lattice if and only if for all $x \in B$ there exists $x' \in B$ such that for all $y \in B$

$$x \leqslant y \iff x \wedge y' = 0,$$

where 0 is some fixed element of B. (This rather tricky exercise in axiomatics gives an extremely useful characterization of Boolean lattices as it eliminates the need to deal with joins and the distributive law. It is interesting that O. Frink's original proof of this result is short and non-computational but relies, unnecessarily, on (ZL).)

7.8 A ring B with identity is called a **Boolean ring** if $x^2 = x$ for all $x \in B$.

(i) Show that the following identities hold in a Boolean ring:

(a) $xy + yx = 0$;   (b) $x + x = 0$;   (c) $xy = yx$.

(ii) Let B be a Boolean algebra and define $+$ and $\cdot$ on B by

$$x + y := (x \wedge y') \vee (x' \wedge y), \quad x \cdot y := x \wedge y.$$

Show that $\langle B; +, \cdot \rangle$ is a Boolean ring.

(iii) Let B be a Boolean ring and define $\vee$ and $\wedge$ on B by

$$x \vee y := x + y + xy, \quad x \wedge y := xy, \quad x' := 1 + x.$$

Show that $\langle B; \vee, \wedge, ', 0, 1 \rangle$ is a Boolean algebra. [Hint. Use Exercise 7.7.]

(iv) Show that the correspondence between Boolean algebras and Boolean rings established in (ii) and (iii) is a bijective one.

7.9 Show that under the correspondence set up in the previous exercise, the Boolean algebra **2** corresponds to the Boolean ring $\mathbb{Z}_2$. Show that the additive group of the Boolean ring corresponding to the Boolean algebra $\mathbf{2}^2$ is isomorphic to the Klein 4-group. (In general, the additive group of the Boolean ring corresponding to $\mathbf{2}^n$ is isomorphic to $\mathbb{Z}_2^n$.)

7.10 Show that an equivalence relation on a Boolean algebra B is a Boolean congruence if and only it is a lattice congruence. Prove that every congruence on a Boolean algebra B is of the form θ_J for some ideal J in B (see Exercise 6.19) and moreover $(a, b) \in \theta_J$ if and only if $(a \wedge b') \vee (a' \wedge b) \in J$.

7.11 Use Theorem 7.12 to find a Boolean term $p(x, y)$ such that F_p coincides with the function $equal : \mathbf{2}^2 \to \mathbf{2}$, where

$$equal\,(x, y) := \begin{cases} 1 & \text{if } x = y, \\ 0 & \text{if } x \neq y. \end{cases}$$

7.12 For any non-empty set, S, the **ternary discriminator** on S is a function $d : S^3 \to S$ given by

$$d(x, y, z) := \begin{cases} z & \text{if } x = y, \\ x & \text{if } x \neq y. \end{cases}$$

(This function plays an important role in universal algebra.)

(i) Give a formula for the ternary discriminator on **2** in terms of $\vee, \wedge, '$ and the function $equal : \mathbf{2}^2 \to \mathbf{2}$ of the previous exercise. Hence give a Boolean term $p(x, y, z)$ such that F_p coincides with the ternary discriminator on **2**.

(ii) Use Theorem 7.12 to find a Boolean term $q(x, y, z)$ such that F_q equals the ternary discriminator on **2**.

(iii) Show via the laws of Boolean algebra that for every Boolean algebra B and all $a, b, c \in B$ we have $p(a, b, c) = q(a, b, c)$.

(iv) Find a 'simple' term r such that F_r agrees with the ternary discriminator on $\mathbf{2}$.

7.13 Design a gate diagram for a majority voting machine with four inputs which allows a current to flow when three or more of the inputs are 1.

7.14 A safe–unsafe decision circuit is required. There are to be three inputs and two outputs, one to operate a green light and the other to operate a red light. The green light should come on if all inputs indicate safety (1), and the red light should come on if any input indicates danger (0). Design the gate diagram.

7.15 The Duality Principle extends to Boolean algebras. Given a statement Φ about Boolean algebras, involving the symbols $\vee, \wedge, ', 0, 1$ and $\leqslant$, indicate the replacements required to produce the dual statement Φ^{∂}. Hence formulate and prove the **Boolean Duality Principle**.

Define what it means for a Boolean term to be in **conjunctive normal form**, or CNF. (This is simply the dual of DNF.)

State the dual of Theorem 7.12 which ensures that every Boolean term is equivalent to a term in CNF.

7.16 For each of the following Boolean terms, p, draw up a truth table for F_p then apply Theorem 7.12 to obtain an equivalent polynomial in DNF. Show that the algorithm given in 7.13 yields the same DNF.

(i) $x \wedge (y \vee z)$;

(ii) $(x \vee y') \wedge z$;

(iii) $(x \vee y') \wedge (y \vee z') \wedge (z \vee x')$;

(iv) $((x \wedge y') \wedge z')' \wedge (x \vee z)$.

7.17 Test each of the following proposed Boolean algebra identities by reducing both sides to DNF:

(i) $(a \wedge b)' \vee (a \wedge c') = (a \wedge (b \vee c')')'$;

(ii) $(a \wedge b) \vee (a \wedge c') \vee (b' \wedge c') = (a \wedge b) \vee (b \vee c)'$;

(iii) $(a' \vee b') \vee (c' \wedge (a \vee b))' = a$.

7.18 The binary operation, $|$, of **inclusive denial** on a Boolean algebra is defined by $x|y := x' \vee y'$. Show that $\vee, \wedge, ', 0$ and 1 can each be defined in terms of | alone. (It is a somewhat more difficult exercise to prove that if $*$ is a binary connective from which each of $\vee$, $\wedge$, $'$, 0 and 1 can be defined, then $*$ is logically equivalent to either inclusive denial or to its dual, $\downarrow$, defined by $x \downarrow y := x' \wedge y'$ and known as **joint denial**.)

7.19 Let B be a Boolean algebra and for $X \subseteq B$ let $[X]$ be the smallest subalgebra of B containing X (cf. Exercise 5.5). Show that

$$[X] = \{\, p(a_1, \ldots, a_n) \mid n \in \mathbb{N}_0, p \in \mathbf{BT}, a_1, \ldots, a_n \in B \,\}.$$

(Note that the case $n = 0$ is included to cover the '0–ary' Boolean terms 0 and 1.)

7.20 Find the subalgebra of the powerset algebra $\wp(\{1,2,3,4,5\})$ generated by $X = \{\,\{1,2\}, \{1,2,3,4\}\,\}$. Draw a diagram of $[X]$ and label each element with an appropriate Boolean term $p(a,b)$ where $a = \{1,2\}$ and $b = \{1,2,3,4\}$. (Compare with Exercise 5.5.)

7.21 (For those who want to get to know LINDA better.)

(i) Check that each axiom of the formal system **L**, as given in 7.15, is a tautology.

(ii) Show that, as claimed in 7.18, the orders defined on $\mathrm{LA}_{\models}$ and $\mathrm{LA}_{\vdash}$ are well defined.

(iii) By following the hints given in 7.18, prove that $[\varphi \vee \psi]$ is the least upper bound of $[\varphi]$ and $[\psi]$ in $\mathrm{LA}_{\vdash}$.

(iv) Let v be a valuation. Prove that the map $f_v \colon \mathrm{LA}_{\vdash} \to \mathbf{2}$, as defined in 7.19, is a well-defined Boolean homomorphism and that every Boolean homomorphism from $\mathrm{LA}_{\vdash}$ to $\mathbf{2}$ arises in this way from some valuation.

8

Representation Theory: the Finite Case

In previous chapters we have introduced various classes of lattices. We have given examples of members of these classes, and described some of their general properties. We now turn our attention to structure theorems. In Chapter 10 we give a concrete representation, as a lattice of sets, of any (bounded) distributive lattice. This chapter deals, less ambitiously, with the finite case, and reveals a very satisfactory correspondence between finite distributive lattices and finite ordered sets. We show that any finite distributive lattice L can be realized as a lattice $\mathcal{O}(P)$ of down-sets built from a suitable subset P of L. En route we investigate, in greater generality than is necessary for our later needs, the problem of finding a subset of a lattice L which, as an ordered set, uniquely determines L. We are thereby able to present some interesting applications of results in Chapter 2. We begin by considering finite Boolean algebras, for which the representation theory is especially simple and easy to motivate.

The representation of finite Boolean algebras

Our archetypal example of a Boolean algebra is $\langle \mathcal{P}(X); \cup, \cap, ', \varnothing, X \rangle$. Any $A \in \mathcal{P}(X)$ is a union of singleton sets $\{x\}$ for $x \in X$. Thus the Boolean algebra $\mathcal{P}(X)$ may be regarded as being built up from the set X. We seek something analogous in an abstract Boolean algebra. Note that in $\mathcal{P}(X)$ the singletons are exactly the elements which cover 0.

8.1 Atoms. Let L be a lattice with least element 0. Then $a \in L$ is called an **atom** if $0 \prec a$. The set of atoms of L is denoted by $\mathcal{A}(L)$.

It may happen that a lattice has no atoms at all. The chain of non-negative real numbers provides an example. Even a Boolean lattice may have no atoms; see Exercise 7.5. However, given any element $a \neq 0$ in a finite lattice there exists an atom x such that $0 \prec x \leqslant a$. In a finite Boolean lattice we can say even more.

8.2 Lemma. *Let B be a finite Boolean lattice. Then, for each $a \in B$,*

$$a = \bigvee \{ x \in \mathcal{A}(B) \mid x \leqslant a \}.$$

Proof. Fix $a \in B$. Let $S = \{x \in \mathcal{A}(B) \mid x \leqslant a\}$. Certainly a is an upper bound for S. Let b be any upper bound for S. To complete the proof we require $a \leqslant b$. Suppose not, so $0 < a \wedge b'$, by 7.3(v). Choose $x \in \mathcal{A}(B)$ such that $0 \prec x \leqslant a \wedge b'$. Then $x \in S$, so $x \leqslant b$. Since also $x \leqslant b'$, we have $x \leqslant b \wedge b' = 0$ ⨑. ∎

8.3 The Representation Theorem for Finite Boolean Algebras. *Let B be a finite Boolean algebra. Then the map*

$$\eta: a \longmapsto \{\, x \in \mathcal{A}(B) \mid x \leqslant a \,\}$$

is an isomorphism of B onto $\wp(X)$, where $X = \mathcal{A}(B)$, with the inverse of η given by $\eta^{-1}(S) = \bigvee S$ for $S \in \wp(X)$.

Proof. We first show that η maps B onto $\wp(X)$. Let $S = \{a_1, \ldots, a_k\}$ be a set of atoms of B and define $a = \bigvee S$. We claim $S = \eta(a)$. Certainly $S \subseteq \eta(a)$. Now let x be any atom such that $x \leqslant a = a_1 \vee \ldots \vee a_k$. For each i, we have $0 \leqslant x \wedge a_i \leqslant x$. Because x is an atom, either $x \wedge a_i = 0$ for all i or there exists j such $x \wedge a_j = x$. In the former case, $x = x \wedge a = (x \wedge a_1) \vee \ldots \vee (x \wedge a_k) = 0$ ⨑. Therefore $x \leqslant a_j$ for some j, which forces $x = a_j$, as a_j and x are atoms. Hence $\eta(a) \subseteq S$, as we wished to show.

Let $a, b \in B$. Then $\eta(a) \subseteq \eta(b)$ implies, by Lemma 8.2, that $a = \bigvee \eta(a) \leqslant \bigvee \eta(b) = b$. It is trivial (by the transitivity of $\leqslant$) that $\eta(a) \subseteq \eta(b)$ whenever $a \leqslant b$, so η is an order-isomorphism. By 5.11(ii) and 7.5, η is an isomorphism of Boolean algebras. ∎

8.4 Corollary. *Let B be a finite lattice. Then the following statements are equivalent:*

  (i) *B is a Boolean lattice;*

 (ii) *$B \cong \wp(\mathcal{A}(B))$;*

(iii) *B is isomorphic to $\mathbf{2}^n$, for some $n \geqslant 0$.*

Further, any finite Boolean lattice has 2^n elements, for some $n \geqslant 0$.

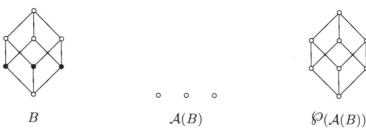

B $\mathcal{A}(B)$ $\wp(\mathcal{A}(B))$

Figure 8.1

8.5 Remarks. To cement the ideas of this section, we illustrate 8.3 in the case that B is a Boolean algebra with 3 atoms; see Figure 8.1.

Corollary 8.4 is a very satisfactory result, but in a sense also a disappointing one. There is little variety among finite Boolean algebras. As we shall see, finite distributive lattices are a much richer class.

Join-irreducible elements

Atoms were the right building blocks for finite Boolean algebras. In order to represent finite distributive lattices we need a more general notion—that of a join-irreducible element.

8.6 Remarks on lattice-building. Our successful representation of a finite Boolean algebra B depended on the existence of a distinguished subset—the atoms of B—from which B could be built. Given any lattice L we might similarly seek a representing set or 'skeleton' P from which to reconstruct L. We should like P to be a subset of L with the following properties:

(i) P is 'small' and readily identifiable;

(ii) L is uniquely determined by the ordered set P.

Even more nebulously, we should also like:

(iii) the process for obtaining L from P is simple to carry out.

Conditions (i) and (ii) pull in opposite directions, since (ii) requires P to be, in some sense, large. Good candidates for sets satisfying (ii) might be those which are join-dense, or (dually) meet-dense, or both. (Recall from 2.34 that P is **join-dense** in L if for every $a \in L$ there exists $A \subseteq P$ such that $a = \bigvee A$.)

The set of atoms of a finite Boolean lattice admirably meets criteria (i), (ii) and (iii). As we have already seen, if the lattice L is not required to be finite and Boolean, there may be too few atoms for the set of atoms to serve as a skeleton. However, alternative candidates for P are available in an arbitrary finite lattice, and, more generally, in a lattice satisfying chain conditions.

8.7 Definitions. Let L be a lattice. An element $x \in L$ is **join-irreducible** if

(i) $x \neq 0$ (in case L has a zero);

(ii) $x = a \vee b$ implies $x = a$ or $x = b$ for all $a, b \in L$.

Condition (ii) is equivalent to the more pictorial

(ii)$'$ $a < x$ and $b < x$ imply $a \vee b < x$ for all $a, b \in L$.

A **meet-irreducible** element is defined dually. We denote the set of join-irreducible elements of L by $\mathcal{J}(L)$ and the set of meet-irreducible elements by $\mathcal{M}(L)$. Each of these sets inherits L's order relation, and will be regarded as an ordered set.

The following lemma compares atoms and join-irreducible elements. It shows that in a Boolean algebra $\mathcal{J}(L)$ coincides with $\mathcal{A}(L)$.

8.8 Lemma.   *Let L be a lattice with least element 0. Then*

   (i) $0 \prec x$ *in L implies $x \in \mathcal{J}(L)$;*

  (ii) *if L is a Boolean lattice, $x \in \mathcal{J}(L)$ implies $0 \prec x$.*

Proof. To prove (i), suppose by way of contradiction that $0 \prec x$ and $x = a \vee b$ with $a < x$ and $b < x$. Since $0 \prec x$, we have $a = b = 0$, whence $x = 0$ ↯.

Now assume L is a Boolean lattice and that $x \in \mathcal{J}(L)$. Suppose $0 \leqslant y < x$; we want $y = 0$. We have

$$x = x \vee y = (x \vee y) \wedge (y' \vee y) = (x \wedge y') \vee y.$$

Figure 8.2

Since x is join-irreducible and $y < x$, we must have $x = x \wedge y'$, whence $x \leqslant y'$. But then $y = x \wedge y \leqslant y' \wedge y = 0$, so $y = 0$. This proves (ii).   ∎

8.9 Examples of join-irreducible elements.

(1) In a chain, every non-zero element is join-irreducible. Thus if L is an n-element chain, then $\mathcal{J}(L)$ is an $(n-1)$-element chain.

(2) In a finite lattice L, an element is join- irreducible if and only if it has exactly one lower cover. This makes $\mathcal{J}(L)$ extremely easy to identify from a diagram of L. Figure 8.3 gives some examples. The join-irreducible elements are shaded.

(3) We next consider the join-irreducible elements in the product $L = L_1 \times L_2$ of lattices L_1 and L_2 each having a least element. First observe that $(x_1, x_2) = (x_1, 0) \vee (0, x_2)$. Thus (x_1, x_2) is not join-irreducible unless either x_1 or x_2 is zero. Further, $x_1 = a_1 \vee b_1$ in L_1 implies $(x_1, 0) = (a_1, 0) \vee (b_1, 0)$. It follows that

$$\mathcal{J}(L) \subseteq (\mathcal{J}(L_1) \times \{0\}) \cup (\{0\} \times \mathcal{J}(L_2)).$$

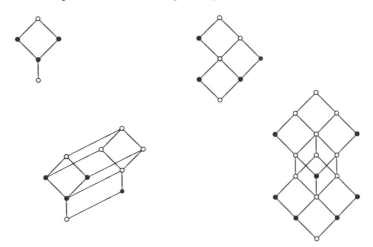

Figure 8.3

It is easily seen that the reverse inclusion also holds. As an ordered set, $\mathcal{J}(L_1 \times L_2)$ is therefore isomorphic to the disjoint union of $\mathcal{J}(L_1)$ and $\mathcal{J}(L_2)$. This example is pursued in 8.22.

(4) Let P be an ordered set. We seek the join-irreducible elements in the lattice $\mathcal{O}(P)$ of down-sets of P. Suppose $x \in P$ and $\downarrow x = U \cup V$, where $U, V \in \mathcal{O}(P)$. Without loss of generality, $x \in U$. But then $\downarrow x \subseteq U$. Since $\downarrow x = U \cup V$ implies $U \subseteq \downarrow x$, we conclude that $\downarrow x = U$. This shows that $\downarrow x \in \mathcal{J}(\mathcal{O}(P))$.

Now suppose P is finite. Any $U \in \mathcal{O}(P)$ is the union of sets $\downarrow x_i$ ($i = 1, \dots, k$), where $x_i \parallel x_j$ for $i \neq j$. Unless $k = 1$, U is not join-irreducible (by 1.19). Hence $\mathcal{J}(\mathcal{O}(P)) = \{ \downarrow x \mid x \in P \}$.

In the last paragraph P must be finite: $\{ q \in \mathbb{Q} \mid q < 0 \}$ is join-irreducible in $\mathcal{O}(\mathbb{Q})$, but is not of the form $\downarrow x$.

(5) It is easily seen that the lattice of open subsets $\mathbb{R}$ (that is, sets which are unions of open intervals in $\mathbb{R}$) has no join-irreducible elements.

Recall from Definition 2.24 that P satisfies the descending chain condition, (DCC), if every descending sequence in P is finite, that is, if $x_1 \geqslant x_2 \geqslant \dots \geqslant x_n \geqslant \dots$, then $x_k = x_{k+1} = \dots$ for some $k \in \mathbb{N}$. Of course, any finite lattice satisfies (DCC).

8.10 Lemma. *Let L be a lattice satisfying* (DCC).

(i) *Suppose $a, b \in L$ and $a \nleqslant b$. Then there exists $x \in \mathcal{J}(L)$ such that $x \leqslant a$ and $x \nleqslant b$.*

(ii) $a = \bigvee \{ x \in \mathcal{J}(L) \mid x \leqslant a \}$ *for all $a \in L$.*

Proof. Let $a \nleqslant b$ and let $S := \{\, x \in L \mid x \leqslant a \ \text{ and } \ x \nleqslant b \,\}$. The set S is non-empty since it contains a. Hence, since L satisfies (DCC), there exists a minimal element x in S (via the dual of Lemma 2.26). We claim that x is join-irreducible. Suppose that $x = c \vee d$ with $c < x$ and $d < x$. By the minimality of x, neither c nor d lies in S. We have $c < x \leqslant a$, so $c \leqslant a$, and similarly $d \leqslant a$. Therefore $c, d \notin S$ implies $c \leqslant b$ and $d \leqslant b$. But then $x = c \vee d \leqslant b$, $\lightning$. Thus $x \in \mathcal{J}(L) \cap S$, which proves (i).

For each $a \in L$ let $T := \{\, x \in \mathcal{J}(L) \mid x \leqslant a \,\}$. Clearly a is an upper bound of T. Let c be an upper bound of T. We claim that $a \leqslant c$. Suppose that $a \nleqslant c$; then $a \wedge c < a$ and hence $a \nleqslant a \wedge c$. By (i) there exists $x \in T$ with $x \nleqslant a \wedge c$. But $x \in T$ implies $x \leqslant a$ (by definition) and $x \leqslant c$ since c is an upper bound of T. Thus x is a lower bound of $\{a, c\}$ and consequently $x \leqslant a \wedge c$ $\lightning$. Hence $a \leqslant c$, as claimed. This proves that $a = \bigvee T$ in L, whence (ii) holds. $\blacksquare$

The remaining results in this section are of independent interest but are not used in the sequel.

8.11 Completion revisited. In Chapter 2 we proved that, for any ordered set P, the Dedekind–MacNeille completion,

$$\mathbf{DM}(P) := \{\, A \subseteq P \mid A^{u\ell} = A \,\},$$

is a complete lattice when ordered by inclusion and that $x \mapsto {\downarrow} x$ is an order-embedding of P into $\mathbf{DM}(P)$. Moreover it was proved in Theorem 2.36 that, if L is a complete lattice and P is a subset of L which is both join-dense and meet-dense in L, then $L \cong \mathbf{DM}(P)$.

Now let L be a lattice with no infinite chains. Since L satisfies both (ACC) and (DCC), it follows by Lemma 8.10 that $\mathcal{J}(L) \cup \mathcal{M}(L)$ is both join-dense and meet-dense in L. Also, by Theorem 2.28, L is complete. Thus we have established the first half of the following result.

8.12 Theorem. *Let L be a lattice with no infinite chains. Then*

$$L \cong \mathbf{DM}(\mathcal{J}(L) \cup \mathcal{M}(L)).$$

Moreover, $\mathcal{J}(L) \cup \mathcal{M}(L)$ is the smallest subset of L which is both join-dense and meet-dense in L.

Proof. It remains to show that, if P is both join-dense and meet-dense in L, then $\mathcal{J}(L) \cup \mathcal{M}(L) \subseteq P$. Let $x \in \mathcal{J}(L)$. Since P is join-dense there exists a subset A of P with $x = \bigvee A$. Hence, by 2.28(i), there is a finite subset F of A with $x = \bigvee F$. Since x is join-irreducible we have $x \in F$ and so $x \in A$. Hence $\mathcal{J}(L) \subseteq A$. By duality, $\mathcal{M}(L) \subseteq A$ too. $\blacksquare$

8.13 A secret revealed. The last theorem is illustrated by the examples which were given when the Dedekind–MacNeille completion was first introduced. Referring back to 2.38(5) and the associated Figure 2.7 we see that in the diagram of each lattice L the set of shaded elements is $\mathcal{J}(L) \cup \mathcal{M}(L)$. Indeed it was via Theorem 8.12 that these examples were constructed in the first place.

8.14 Stocktaking. In 8.6 we put forward three properties we would wish a representing subset P for a lattice L to possess. We now see that, so long as L has no infinite chains, the ordered set $P = \mathcal{J}(L) \cup \mathcal{M}(L)$ has the first two of these properties. However, in general we have to construct a Dedekind–MacNeille completion to recapture L and it is debatable whether this process could be described as simple. Nonetheless a representation for arbitrary finite lattices can be developed from this starting point, and results about finite lattices derived from corresponding results on finite ordered sets. We do not pursue this further.

In the next section we show that a finite *distributive* lattice, L, is determined by the ordered set $\mathcal{J}(L)$ (or $\mathcal{M}(L)$) alone. More significantly, the procedure for recapturing L from $\mathcal{J}(L)$ is relatively simple: it turns out that $L \cong \mathcal{O}(\mathcal{J}(L))$. The resulting representation theory provides a very powerful tool for studying finite distributive lattices.

The representation of finite distributive lattices

A distributive finite lattice L gives rise to an ordered set $J(L) = P$, and this in turn yields a lattice $\mathcal{O}(P)$. By analogy with Theorem 8.3, we might conjecture that $L \cong \mathcal{O}(P)$. Birkhoff's Representation Theorem, proved in 8.17, asserts that this is indeed the case. Before tackling the theorem, we look at some examples.

8.15 Examples and remarks. Figure 8.4 shows L, $\mathcal{J}(L)$ and $\mathcal{O}(\mathcal{J}(L))$ for some small lattices L (cf. 1.20). Note that $L \cong \mathcal{O}(\mathcal{J}(L))$ in cases (i) and (ii), but not in cases (iii) and (iv), and that, of the four lattices L, just the first two are distributive.

Since $\mathcal{O}(\mathcal{J}(L))$ is always distributive, we cannot have $L \cong \mathcal{O}(\mathcal{J}(L))$ unless L is distributive. Indeed, this observation provides an alternative to the $\mathbf{M_3}$–$\mathbf{N_5}$ Theorem for establishing non-distributivity; see 6.11.

None of our results on join-irreducible elements in the preceding section brought in the distributive law. The next result, which we use in the proof of 8.17, does.

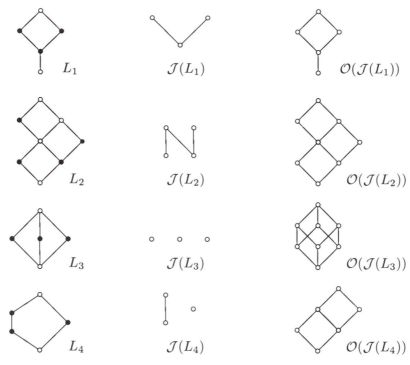

Figure 8.4

8.16 Lemma.   *Let L be a distributive lattice and let $x \in L$, with $x \neq 0$ in case L has a zero. Then the following are equivalent:*

(i) *x is join-irreducible;*

(ii) *if $a, b \in L$ and $x \leqslant a \vee b$ then $x \leqslant a$ or $x \leqslant b$;*

(iii) *for any $k \in \mathbb{N}$, if $a_1, \ldots, a_k \in L$ and $x \leqslant a_1 \vee \ldots \vee a_k$ then $x \leqslant a_i$ for some i ($1 \leqslant i \leqslant k$).*

Proof. To prove that (i) implies (ii), we assume that $x \in \mathcal{J}(L)$ and that $a, b \in L$ are such that $x \leqslant a \vee b$. We have

$$x = x \wedge (a \vee b) \qquad \text{(since } x \leqslant a \vee b)$$
$$= (x \wedge a) \vee (x \wedge b) \qquad \text{(since } L \text{ is distributive).}$$

Because x is join-irreducible, $x = x \wedge a$ or $x = x \wedge b$. Hence $x \leqslant a$ or $x \leqslant b$, as required.

That (ii) implies (iii) is proved by induction on k; the case $k = 1$ is trivial and (ii) gets the induction started at $k = 2$.

It is trivial that (iii) implies (ii), so it only remains to deduce (i) from (ii). Suppose (ii) holds and that $x = a \vee b$. Then certainly $x \leqslant a \vee b$, so $x \leqslant a$ or $x \leqslant b$. But $x = a \vee b$ forces $x \geqslant a$ and $x \geqslant b$. Hence $x = a$ or $x = b$. ∎

8.17 Birkhoff's Representation Theorem for Finite Distributive Lattices. Let L be a finite distributive lattice. Then the map $\eta \colon L \to \mathcal{O}(\mathcal{J}(L))$ defined by

$$\eta(a) = \{\, x \in \mathcal{J}(L) \mid x \leqslant a \,\} \quad (= \mathcal{J}(L) \cap {\downarrow} a)$$

is an isomorphism of L onto $\mathcal{O}(\mathcal{J}(L))$.

Proof. It is immediate that $\eta(a) \in \mathcal{O}(\mathcal{J}(L))$ (since $\leqslant$ is transitive). By Proposition 5.11, it remains only to show that η is an order-isomorphism.

To prove that $a \leqslant b$ implies $\eta(a) \subseteq \eta(b)$, use 2.9(v). To prove that $\eta(a) \subseteq \eta(b)$ implies $a \leqslant b$, use Lemma 8.10 to obtain

$$a = \bigvee \eta(a) \leqslant \bigvee \eta(b) = b.$$

Finally, to prove that η is onto, let $U \in \mathcal{O}(\mathcal{J}(L))$ and write $U = \{a_1, \ldots, a_k\}$. Define a to be $a_1 \vee \ldots \vee a_k$. We claim $U = \eta(a)$. To prove this, first let $x \in U$, so $x = a_i$ for some i. Then x is join-irreducible and $x \leqslant a$, hence $x \in \eta(a)$. In the reverse direction, suppose $x \in \eta(a)$. Then $x \leqslant a = a_1 \vee \ldots \vee a_k$ and Lemma 8.16 implies $x \leqslant a_i$ for some i. Since U is a down-set and $a_i \in U$, we have $x \in U$. ∎

8.18 Corollary. *Let L be a finite lattice. Then the following statements are equivalent:*

  (i) *L is distributive;*

 (ii) *$L \cong \mathcal{O}(\mathcal{J}(L))$;*

(iii) *L is isomorphic to a lattice of sets;*

(iv) *L is isomorphic to a sublattice of $\mathbf{2}^n$ for some $n \geqslant 0$.*

Duality between finite distributive lattices and finite ordered sets

Our first result in this section is a natural companion to Theorem 8.17.

8.19 Theorem. *Suppose P is a finite ordered set. Then the map $\varepsilon \colon x \mapsto {\downarrow} x$ is an order-isomorphism from P onto $\mathcal{J}(\mathcal{O}(P))$.*

Proof. Lemma 1.21 implies that ε is an order-embedding of P into $\mathcal{O}(P)$ and Example 8.9(4) shows that the image of ε is $\mathcal{J}(\mathcal{O}(P))$. ∎

8.20 Finite distributive lattices and ordered sets in partnership.   We denote by $\mathbf{D}_F$ the class of all finite distributive lattices and by $\mathbf{P}_F$ the class of all finite ordered sets. Theorems 8.17 and 8.19 assert that

$$L \cong \mathcal{O}(\mathcal{J}(L)) \quad \text{and} \quad P \cong \mathcal{J}(\mathcal{O}(P))$$

for all $L \in \mathbf{D}_F$ and $P \in \mathbf{P}_F$. We call $\mathcal{J}(L)$ the **dual** of L and $\mathcal{O}(P)$ the **dual** of P. (The use of the word dual here should of course not be confused with that in 1.17.)

By identifying a finite distributive lattice L with the isomorphic lattice $\mathcal{O}(\mathcal{J}(L))$ of down-sets of $\mathcal{J}(L)$, we may regard $\mathbf{D}_F$ as consisting of the concrete lattices $\mathcal{O}(P)$, for $P \in \mathbf{P}_F$, rather than as abstract objects satisfying certain identities.

Up to isomorphism, we have a one-to-one correspondence

$$\mathcal{O}(P) = L \quad \underset{\longleftarrow}{\overrightarrow{\hspace{2cm}}} \quad P = \mathcal{J}(L)$$

for $L \in \mathbf{D}_F$ and $P \in \mathbf{P}_F$. Describing P, given L, is entirely straight-forward. Those who worked through Exercise 1.10 will appreciate that describing L, given P, can be laborious, even when P is quite small. Mi-crocomputer programs have recently been devised for drawing $\mathcal{O}(P)$ for a given finite ordered set P. These are viable only so long as $\mathcal{O}(P)$ re-mains reasonably small (of the order of hundreds). It possible for $|\mathcal{O}(P)|$ to grow extremely fast as P increases, so that the problem of determin-ing $\mathcal{O}(P)$ from P becomes intractable, even with computer assistance. The example $P = \wp(X)$, for $|X| = 1, 2, 3, \dots$, is instructive. Figure 8.5 shows $\wp(X)$ and $\mathcal{O}(\wp(X))$ for $|X| = 3$ and the accompanying table $|\wp(X)|$ and $|\mathcal{O}(\wp(X))|$ for $|X| \leqslant 7$. The size of $\mathcal{O}(\wp(X))$ for $|X| = 8$ remains elusive.

The above observations show that the dual of a finite distributive lattice is generally much smaller and less complex than the lattice itself. This means that lattice problems concerning $\mathbf{D}_F$ are likely to become simpler when translated into problems about $\mathbf{P}_F$. We may regard the maps $L \mapsto \mathcal{J}(L)$ and $P \mapsto \mathcal{O}(P)$ as playing a role analogous to that of the logarithm and exponential functions (and this analogy is strength-ened by 8.22(iii)).

Special properties of a finite distributive lattice are reflected in spe-cial properties of its dual. The following lemma provides an elementary example. Its proof is left as an easy deduction from earlier results.

$\|X\|$	$\|\wp(X)\|$	$\|\mathcal{O}(\wp(X))\|$
1	2	3
2	4	6
3	8	20
4	16	168
5	32	7581
6	64	7828354
7	128	2414682040998

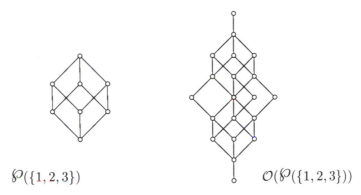

$\wp(\{1,2,3\})$ $\qquad\qquad\qquad$ $\mathcal{O}(\wp(\{1,2,3\}))$

Figure 8.5

8.21 Lemma. *Let $L = \mathcal{O}(P)$ be a finite distributive lattice. Then*

(i) *L is a Boolean lattice if and only if P is an antichain; $\mathcal{O}(\overline{\mathbf{n}}) = \mathbf{2}^n$.*

(ii) *L is a chain if and only if P is a chain; $\mathcal{O}(\mathbf{n}) = \mathbf{n+1}$.*

Our next result serves a double purpose. It illustrates the way in which lattice constructs in $\mathbf{D}_F$ correspond to ordered set constructs in $\mathbf{P}_F$. At the same time, it assists in the analysis and interpretation of the relation between complex lattices and their duals.

8.22 Theorem. *Let P be a finite ordered set. Then*

(i) *$\mathcal{O}(P)^\partial \cong \mathcal{O}(P^\partial)$;*

(ii) *$\mathcal{O}(P \oplus \mathbf{1}) \cong \mathcal{O}(P) \oplus \mathbf{1}$ and $\mathcal{O}(\mathbf{1} \oplus P) \cong \mathbf{1} \oplus \mathcal{O}(P)$;*

(iii) *if $P = P_1 \,\dot\cup\, P_2$, then $\mathcal{O}(P) \cong \mathcal{O}(P_1) \times \mathcal{O}(P_2)$.*

Proof. (i) The required isomorphism is given by $U \mapsto P \smallsetminus U$ for $U \in \mathcal{O}(P)$; see 1.19.

(ii) The down-sets of $P \oplus \mathbf{1}$ are the down-sets of P together with $P \oplus \mathbf{1}$ itself. The down-sets of $\mathbf{1} \oplus P$ are the empty set and all down-sets of

P with the least element of $\mathbf{1} \oplus P$ adjoined. The required isomorphisms are now easily set up.

(iii) It is easily verified that the map $U \mapsto (U \cap P_1, U \cap P_2)$, for $U \in \mathcal{O}(P)$, defines an isomorphism from $\mathcal{O}(P_1 \dot\cup P_2)$ to $\mathcal{O}(P_1) \times \mathcal{O}(P_2)$. Example 8.9(4) gives an alternative proof. ∎

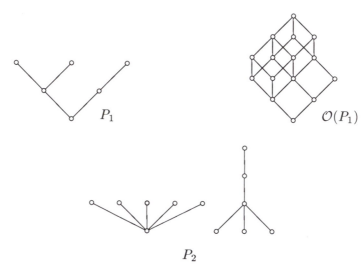

P_1 $\mathcal{O}(P_1)$

P_2

Figure 8.6

8.23 Examples. Consider Figures 8.6 and 8.7.

(1) The ordered set P_1 may be thought of as $\mathbf{1} \oplus ((\mathbf{1} \oplus \overline{\mathbf{2}}) \dot\cup \mathbf{2})$. By 8.22(ii) and (iii), we see that the corresponding lattice $\mathcal{O}(P_1)$ is isomorphic to $\mathbf{1} \oplus ((\mathbf{1} \oplus \mathbf{2}^2) \times \mathbf{3})$.

(2) By repeated use of Theorem 8.22 we see that the lattice dual to P_2 is isomorphic to $(\mathbf{1} \oplus \mathbf{2}^5) \times (\mathbf{2}^3 \oplus \mathbf{3})$. This lattice is too complicated to draw effectively, but we know at least that its size is $(1 + 2^5) \times (2^3 + 3) = 363$.

(3) Consider the lattice L_1. The ordered set $\mathcal{J}(L_1)$ is shown alongside. We have $\mathcal{O}(\mathcal{J}(L_1)) \cong \mathbf{2} \times (\mathbf{1} \oplus \mathbf{2}^2)$, which has 10 elements. We deduce that L_1 is not isomorphic to $\mathcal{O}(\mathcal{J}(L_1))$, so that L_1 is not distributive.

(4) Now consider L_2. We cannot immediately describe $\mathcal{O}(\mathcal{J}(L_2))$. However we can see that $L_2 \cong L_2^\partial$ yet $\mathcal{J}(L_2)$ is not isomorphic to its order dual. Hence, by Theorem 8.22(i), L_2 cannot be isomorphic to $\mathcal{O}(\mathcal{J}(L_2))$ and consequently is not distributive.

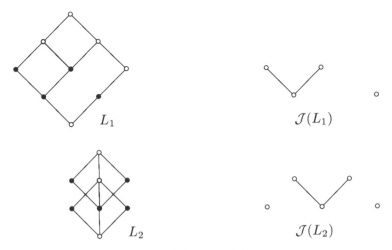

Figure 8.7

Our discussion of the partnership between $\mathbf{D}_F$ and $\mathbf{P}_F$ would be seriously incomplete were we not to consider structure-preserving maps (recall 1.16). Theorem 8.24 sets up a one-to-one correspondence between $\{0,1\}$-homomorphisms from $\mathcal{O}(P)$ to $\mathcal{O}(Q)$ and order-preserving maps from Q to P, for $P, Q \in \mathbf{P}_F$ (note the reversal of the direction). This theorem is harder to formulate and to prove than any of our preceding results on duals. Some of the difficulty stems from our hitherto admirable choice of the join-irreducible elements as the basis for our representation theory. In Chapter 10 we remove the finiteness restrictions under which we currently working. We shall then obtain a more natural version of Theorem 8.24, as a special case of Theorem 10.26. In the meantime we recommend only to the most intrepid readers the exercise of proving Theorem 8.24 directly.

8.24 Theorem. *Let P and Q be finite ordered sets and let $L = \mathcal{O}(P)$ and $K = \mathcal{O}(Q)$.*

Given a $\{0,1\}$-homomorphism $f : L \to K$, there is an associated order-preserving map $\varphi_f : Q \to P$ defined by

$$\varphi_f(y) = \min\{\, x \in P \mid y \in f(\downarrow x) \,\}$$

for all $y \in Q$.

Given an order-preserving map $\varphi : Q \to P$, there is an associated $\{0,1\}$-homomorphism $f_\varphi : L \to K$ defined by

$$f_\varphi(a) = \varphi^{-1}(a) \text{ for all } a \in L.$$

Equivalently,

$$\varphi(y) \in a \text{ if and only if } y \in f_\varphi(a) \quad \text{for all } a \in L, y \in Q.$$

The maps $f \mapsto \varphi_f$ and $\varphi \mapsto \varphi_f$ establish a one-to-one correspondence between $\{0,1\}$-homomorphisms from L to K and order-preserving maps Q to P.

Further,

(i) f is one-to-one if and only if φ_f is onto;

(ii) f is onto if and only if φ_f is an order-embedding.

8.25 Example. Figure 8.8 shows an order-preserving map and the associated $\{0,1\}$-homomorphism. The image of f is shaded.

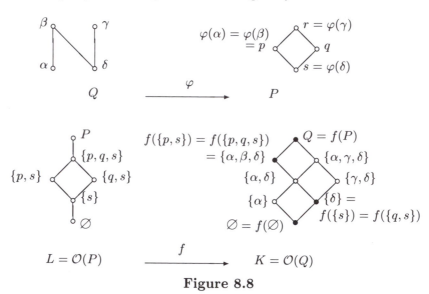

Figure 8.8

8.26 Stocktaking. Combining 8.20 and 8.24, we have a correspondence between $\mathbf{D}_F + \{0,1\}$-homomorphisms and $\mathbf{P}_F + $ order-preserving maps which establishes what in technical parlance is called a **duality** or a **dual equivalence of categories**. In Chapter 10 we specify more explicitly what we mean by a duality, in the context of the representation of all bounded distributive lattices. The import of the duality in the finite case should already be clear: statements about finite distributive lattices can be translated into statements about finite ordered sets, and vice versa.

We can now see that our two uses of the word 'dual' have an underlying commonality. If, in an ordered set P, we think of $x \leqslant y$ as representing an 'arrow' from x to y, then P^∂ is obtained by reversing the arrows. Similarly, for $L, K \in \mathbf{D}_F$, a $\{0,1\}$-homomorphism $f: L \to K$ provides an 'arrow' from L to K, and Theorem 8.24 shows that when we pass from $\mathbf{D}_F$ to $\mathbf{P}_F$ the arrows again reverse. Category theory is exactly the tool needed to formalize this hand waving. The step up into the wide blue, category-theoretic, yonder is not a large one but is beyond the scope of our work here.

Exercises

Exercises from the text. Prove Theorem 8.22(iii) from 8.17, 8.19 and 8.9(3) (see 8.9(3)).

8.1   Consider the lattices in Figure 8.9.

 (i) Draw labelled diagrams of the ordered sets $\mathcal{J}(L)$ and $\mathcal{M}(L)$ for each of the lattices.

 (ii) Draw a labelled diagram of $\mathcal{O}(\mathcal{J}(L))$ in each case and comment on your results.

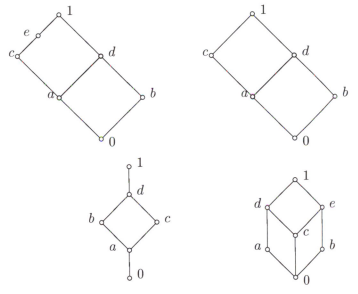

Figure 8.9

8.2   Use Theorem 8.17 to show that the lattices in Figure 8.10 are not distributive.

8.3   Verify Theorem 8.19 for the ordered sets given in Figure 8.11.

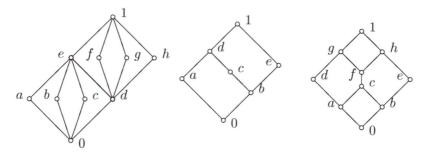

Figure 8.10

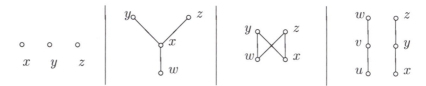

Figure 8.11

8.4 Give counterexamples to each of the following statements.

   (i) If L and K are finite lattices and $\mathcal{J}(L)$ and $\mathcal{J}(K)$ are order-isomorphic, then L and K are isomorphic.

   (ii) If L is a distributive lattice, then $L \cong \mathcal{O}(\mathcal{J}(L))$. [Hint. Consider a suitable infinite chain.]

   (iii) If L is a finite distributive lattice and $\mathcal{J}(L)$ is a lattice, then $\mathcal{J}(L)$ is a sublattice of L.

   (iv) If L is a finite lattice, then $\mathcal{J}(L) \cong \mathcal{M}(L)$.

8.5 Let L be a finite distributive lattice. Prove by the steps below that $\mathcal{J}(L) \cong \mathcal{M}(L)$. (An alternative proof, by duality, is indicated in Exercise 8.6.)

   (i) Let $x \in \mathcal{J}(L)$. Show that there exists $x' \in L$ such that ${\downarrow} x' = L {\smallsetminus} {\uparrow} x$. [Hint. Let $x' := \bigvee(L {\smallsetminus} {\uparrow} x)$ and then use Lemma 8.16 to show that $x' \not\geq x$.]

   (ii) Show that for all $x \in \mathcal{J}(L)$ the element x' defined in (i) is meet-irreducible.

   (iii) Prove that the map $\varphi \colon \mathcal{J}(L) \to \mathcal{M}(L)$, given by $\varphi(x) = x'$ for all $x \in \mathcal{J}(L)$, is an order-isomorphism. [Hint. Recall from Lemma 1.21 that $x \leqslant y$ if and only if ${\uparrow} x \supseteq {\uparrow} y$. When proving that φ maps onto $\mathcal{M}(L)$, use the dual of (i) and (ii).]

8.6 Let P be a finite ordered set.

    (i) Show that a down-set U is meet-irreducible in $\mathcal{O}(P)$ if and only if it is of the form $P \smallsetminus {\uparrow}x$ for some $x \in P$.

    (ii) Use (i) to show that P is order-isomorphic to $\mathcal{M}(\mathcal{O}(P))$.

    (iii) Conclude that $\mathcal{J}(\mathcal{O}(P)) \cong \mathcal{M}(\mathcal{O}(P))$. Deduce that $\mathcal{J}(L)$ is order-isomorphic to $\mathcal{M}(L)$ for any finite distributive lattice L.

8.7 Prove that the length of a finite distributive lattice L equals $|\mathcal{J}(L)|$. [Hint. Use Exercise 1.9.]

8.8 Let L be a finite distributive lattice. Prove that the width of $\mathcal{J}(L)$ equals the least $n \in \mathbb{N}$ such that L can be embedded into a product of n chains. [Hint. Use the duality between $\mathbf{D}_F$ and $\mathbf{P}_F$ to reinterpret Dilworth's Theorem—see Exercise 1.23.]

8.9 Let P be an ordered set with $\top$ and Q an ordered set with $\bot$. The **vertical sum** of P and Q, denoted $P \overline{\oplus} Q$, is obtained from the linear sum $P \oplus Q$ by identifying the top of P with the bottom of Q.

    (i) Let P and Q be finite ordered sets. Show that

$$\mathcal{O}(P \oplus Q) \cong \mathcal{O}(P)\overline{\oplus}\mathcal{O}(Q).$$

    (ii) Derive Theorem 8.22 (ii) from (i) above.

8.10 Use the method illustrated in Examples 8.23, along with the result of the previous exercise, to describe $\mathcal{O}(P)$ for each of the ordered sets P in Figure 8.12. Give the cardinality of $\mathcal{O}(P)$ in each case.

8.11 Use the duality between finite distributive lattices and finite ordered sets to answer the question posed in Exercise 5.12(ii). (First prove that the lattice is distributive!)

8.12 Let L be $\langle \mathbb{N}_0; \mathrm{lcm}, \gcd \rangle$.

    (i) Draw diagrams of ${\downarrow}4$ and ${\downarrow}12$.

    (ii) Show that for every $k \geqslant 1$ there exists $n_k \in \mathbb{N}$ such that ${\downarrow}n_k \cong \wp(\{1, 2, \ldots, k\})$.

    (iii) Deduce that every finite distributive lattice can be embedded in L.

    (iv) Give an example of a countable distributive lattice which cannot be embedded into L.

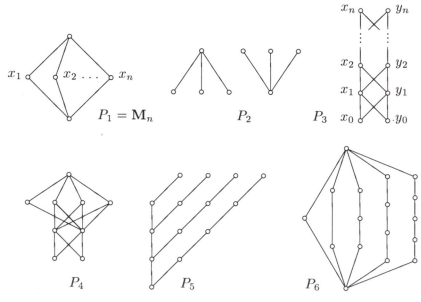

Figure 8.12

8.13 Let L be a finite distributive lattice. Prove that there exists a finite Boolean lattice B and an embedding $\eta \colon L \to B$ such that η is a $\{0,1\}$-homomorphism. Show further that, if $|L| = n$, then B can be chosen so that $|B| \leqslant 2^{n-1}$.

8.14 Consider the (distributive) lattice $\langle \mathbb{N}_0; \mathrm{lcm}, \gcd \rangle$ and let $n \in \mathbb{N}$.

  (i) Show that $\mathcal{J}(\mathbb{N}) = \{\, m \in \mathbb{N} \mid m \text{ is a power of a prime}\,\}$ and hence describe $\mathcal{J}(\!\downarrow\! n)$.

  (ii) Let $n = p_1^{k_1} \ldots p_s^{k_s}$ with the p_i pairwise distinct primes. Show that
  $$\downarrow\! n \cong (\mathbf{k}_1 \oplus \mathbf{1}) \times \cdots \times (\mathbf{k}_s \oplus \mathbf{1}).$$

8.15 Prove that the lattice $\operatorname{Sub}\mathbb{Z}_n$ of subgroups of the cyclic group $\mathbb{Z}_n = \{0, 1, \ldots, n-1\}$, under addition modulo n, is a finite product of finite chains. [Hint. First prove that $\operatorname{Sub}\mathbb{Z}_n$ is isomorphic to the dual of the principal ideal $\downarrow\! n$ in $\langle \mathbb{N}_0; \mathrm{lcm}, \gcd \rangle$, then prove that $(\!\downarrow\! n)^\partial \cong \downarrow\! n$ and finally apply Exercise 8.14.]

8.16 Let $m_1, m_2, \ldots, m_s \in \mathbb{N}$ with $m_i > 1$ for all i. Use the duality between $\mathbf{D}_F$ and $\mathbf{P}_F$ to prove that
$$\mathbf{m}_1 \times \cdots \times \mathbf{m}_s \cong \mathbf{2}^t \implies m_1 = \cdots = m_s = 2 \text{ and } s = t.$$

8.17 Use Exercises 8.15 and 8.16 along with 6.6(4) to characterize those groups G such that $\mathrm{Sub}\,G$ is (i) a finite chain, (ii) isomorphic to $\wp(X)$ for some finite set X.

8.18 Recall from 1.28 that, if P and Q are ordered sets, then $Q^{\langle P\rangle}$, or alternatively $\langle P \to Q\rangle$, denotes the set of order-preserving maps from P to Q with the pointwise order.

  (i) Let L be a lattice and P an ordered set. Show that $L^{\langle P\rangle}$ is a sublattice of L^P and hence is distributive whenever L is.

  (ii) Prove that a subset U of an ordered set Q is an up-set if and only if its characteristic function $\chi_U \colon Q \to \mathbf{2}$ is order-preserving. (Here $\chi_U(x) = 1$ if $x \in U$ and $\chi_U(x) = 0$ if $x \notin U$.) Show further that $\mathcal{O}(Q) \cong \langle Q \to \mathbf{2}\rangle^{\partial}$.

  (iii) Prove that $\langle Q \to \mathbf{2}\rangle^{\partial} \cong \langle Q^{\partial} \to \mathbf{2}\rangle$.

  (iv) Use (ii), (iii) and Exercise 1.15 to show that, for all ordered sets P and Q,

$$\mathcal{O}(Q)^{\langle P\rangle} \cong \mathcal{O}(P^{\partial} \times Q).$$

  (v) Conclude that, if $L \in \mathbf{D}_F$ and $P \in \mathbf{P}_F$, then

$$\mathcal{J}(L^{\langle P\rangle}) \cong P^{\partial} \times \mathcal{J}(L).$$

  (vi) Hence draw diagrams of $\mathbf{2}^{\langle \mathbf{4}\rangle}$ and $\mathbf{4}^{\langle \mathbf{2}\rangle}$.

8.19 Let L be a finite distributive lattice with $|L| > 1$ and let X denote the set of all $\{0,1\}$-homomorphisms from L to $\mathbf{2}$ ordered pointwise.

  (i) Let $x \in \mathcal{J}(L)$ and define $f_x \colon L \to \mathbf{2}$ by

$$(\forall a \in L)\, f_x(a) = \begin{cases} 1 & \text{if } a \geqslant x, \\ 0 & \text{if } a \not\geqslant x, \end{cases}$$

that is, f_x is the characteristic function of $\uparrow x$. Show that f_x is a $\{0,1\}$-homomorphism.

  (ii) Prove that the map $\varepsilon \colon \mathcal{J}(L) \to X^{\partial}$, defined by $\varepsilon(x) = f_x$ for all $x \in \mathcal{J}(L)$, is an order-isomorphism from $\mathcal{J}(L)$ onto X^{∂} whose inverse $\eta \colon X^{\partial} \to \mathcal{J}(L)$ is given by

$$\eta(f) = \bigwedge \{\, a \in L \mid f(a) = 1 \,\} \text{ for all } f \in X.$$

8.20 Consider the distributive lattices shown in Figure 8.13.

  (i) Is it possible to find an onto $\{0,1\}$-homomorphism $f\colon L \to K$ where $L = L_1$ and $K = L_2$? (Use Theorem 8.24(ii) to justify your answer.)

  (ii) Repeat (i) with $L = L_1$ and $K = L_3$.

  (iii) Repeat (i) with $L = L_4$ and $K = L_5$.

  (iv) Repeat (i) with $L = L_6$ and $K = L_7$.

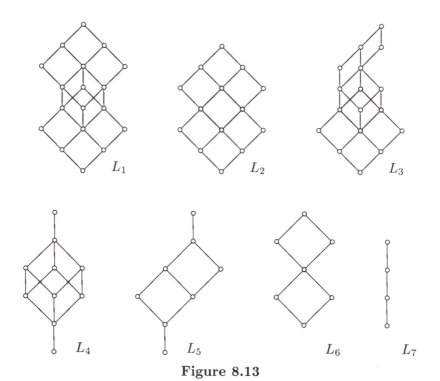

Figure 8.13

8.21 Prove that if L is a finite distributive lattice then $\operatorname{Con} L$ is isomorphic to $\mathbf{2}^n$ where $n = |\mathcal{J}(L)|$. [Hint. Either (a) prove this indirectly via Theorem 8.24 and the fact that there is a bijection between congruences on L and the kernels of homomorphisms with domain L (see Lemma 5.19 and Theorem 5.20), or (b) define $f\colon \wp(\mathcal{J}(L)) \to (\operatorname{Con} L)^{\partial}$ by $f(Y) = \theta_Y$ for all $Y \subseteq \mathcal{J}(L)$, where

$$a \equiv b \,(\mathrm{mod}\, \theta_Y) \quad \Longleftrightarrow \quad {\downarrow}a \cap Y = {\downarrow}b \cap Y \;\; (a,b \in L),$$

and show that f is an order-isomorphism.]

8.22 A lattice L with 0 is said to be **pseudocomplemented** if, for each $a \in L$, there exists an element $a^* \in L$ such that, for all $b \in L$,

$$a \wedge b = 0 \quad \Longleftrightarrow \quad b \leqslant a^*,$$

that is, $a^* = \max\{\, b \in L \mid a \wedge b = 0 \,\}$.

(i) Show that any Boolean lattice is pseudocomplemented.

(ii) Show that any bounded chain is pseudocomplemented.

(iii) Show that any finite distributive lattice is pseudocomplemented.

(iv) Prove that, if $P \in \mathbf{P}_F$, then for each $a \in L := \mathcal{O}(P)$ we have $a^* = P \smallsetminus {\uparrow}a$. (Here ${\uparrow}a$ is calculated in P not in L.)

(v) Give an example of a bounded distributive lattice which is not pseudocomplemented.

8.23 Let P and Q be finite ordered sets and let $L = \mathcal{O}(P)$ and $K = \mathcal{O}(Q)$ be the corresponding finite distributive lattices. Assume that $f : L \to K$ is a $\{0, 1\}$-homomorphism and that $\varphi : Q \to P$ is the dual of f. (See Theorem 8.24.)

Show that the following are equivalent:

(i) $f(a^*) = (f(a))^*$ for all $a \in L$ (where a^* is as given in Exercise 8.22);

(ii) $\mathrm{Min}\,(\varphi(y)) = \varphi(\mathrm{Min}\,(y))$ for all $y \in Q$ (where $\mathrm{Min}\,(z)$ denotes the set of minimal elements in ${\downarrow}z$).

9
Ideals and Filters

This chapter develops the basic theory of ideals and filters in lattices. Ideals are of fundamental importance in algebra. Filters, the order duals of lattice ideals, have a variety of applications in logic and topology.

Ideals—specifically prime ideals—form the basis for our representation theory in Chapter 10. Join-irreducible elements served admirably as building blocks for finite distributive lattices, but we need an alternative if we are to remove the finiteness restriction. Example 8.9(5) shows that an infinite distributive lattice may have no join-irreducible elements at all.

Ideals and filters

Ideals have already appeared in the exercises. We now introduce them 'officially', and add a bevy of further definitions.

9.1 Definitions. Let L be a lattice. A non-empty subset J of L is called an **ideal** if

(i) $a, b \in J$ imply $a \vee b \in J$,

(ii) $a \in L$, $b \in J$ and $a \leqslant b$ imply $a \in J$.

Thus an ideal is a non-empty down-set closed under join. We have spelt out the definition to draw a parallel between a lattice ideal and an ideal in a ring. Exercise 9.4 shows the connection to be stronger than just an analogy.

A dual ideal is called a **filter**. Specifically, a non-empty subset G of L is called a **filter** if

(i) $a, b \in G$ imply $a \wedge b \in G$,

(ii) $a \in L$, $b \in G$ and $a \geqslant b$ imply $a \in G$.

The set of all ideals (filters) of L is denoted by $\mathcal{I}(L)$ (by $\mathcal{F}(L)$). Each of these, with the empty set added, is a topped $\bigcap$-structure and hence a complete lattice.

An ideal or filter is called **proper** if it does not coincide with L. It is a very easy exercise to show that

(i) an ideal J of a lattice with 1 is proper if and only if $1 \notin J$;

(ii) a filter G of a lattice with 0 is proper if and only if $0 \notin G$.

For each $a \in L$, the set $\downarrow a$ is an ideal (the **principal ideal** generated by a); dually, $\uparrow a$ is a **principal filter**. Given any non-empty subset A of L there is a smallest ideal containing A, namely

$$(A] := \{\, a \in L \mid a \leqslant \bigvee T \text{ for some finite subset } T \text{ of } A \,\}.$$

9.2 Examples.

(1) In a finite lattice, every ideal or filter is principal: the ideal J is $\downarrow\!\bigvee J$. (See Exercise 2.12(iii).)

(2) Let L and K be bounded lattices and $f \colon L \to K$ a $\{0,1\}$-homomorphism. Then $f^{-1}(0)$ is an ideal and $f^{-1}(1)$ is a filter in L.

(3) The following are ideals in $\wp(X)$:

   (a) all subsets not containing a fixed element of X;

   (b) all finite subsets (this ideal is non-principal if X is infinite).

(4) Let $(X; \mathcal{T})$ be a topological space and let $x \in X$. Then the set $\{\, V \subseteq X \mid (\exists U \in \mathcal{T})\, x \in U \subseteq V \,\}$ is a filter in $\wp(X)$. Convergence in a topological space can be elegantly formulated in terms of such neighbourhood filters; see [2].

Prime ideals, maximal ideals and ultrafilters

This section introduces ideals and filters of special types.

9.3 Definitions. Let L be a lattice and J a proper ideal in L. Then J is said to be **prime** if $a, b \in L$ and $a \wedge b \in J$ imply $a \in J$ or $b \in J$. The set of prime ideals of L is denoted $\mathcal{I}_p(L)$. It is ordered by set inclusion. A **prime filter** is defined dually and the set of prime filters is denoted by $\mathcal{F}_p(L)$.

A subset J of a lattice L is a prime ideal if and only if $L \setminus J$ is a prime filter—a simple exercise. Thus it is easy to switch between $\mathcal{I}_p(L)$ and $\mathcal{F}_p(L)$. In the sequel we work predominantly with prime ideals. The next two results provide evidence that these may act as a substitute for join-irreducible elements in the representation of distributive lattices.

9.4 Proposition. *Let L be a finite distributive lattice and let $\emptyset \neq J \subseteq L$. Then J is a prime ideal if and only if $J = L \setminus \uparrow x$ for some $x \in \mathcal{J}(L)$. Further, $\mathcal{J}(L) \cong \mathcal{I}_p(L)$, via the map $\varphi \colon x \mapsto L \setminus \uparrow x$.*

Proof. Lemma 8.16 asserts that $\uparrow x$ is a prime filter if and only if $x \in \mathcal{J}(L)$. Hence, taking complements, we have

$$\mathcal{I}_p(L) = \{\, L \setminus \uparrow x \mid x \in \mathcal{J}(L) \,\}.$$

We now know that φ maps $J(L)$ onto $\mathcal{I}_p(L)$. By the dual of 1.21, $x \leqslant y$ if and only if $\uparrow x \supseteq \uparrow y$, so φ is an order-embedding. ∎

The following corollary reformulates Lemma 8.10(i)—a critical step in the proof of Birkhoff's Representation Theorem.

9.5 Corollary. *Let L be a finite distributive lattice and let $a \not\leqslant b$ in L. Then there exists $I \in \mathcal{I}_p(L)$ such that $a \notin I$ and $b \in I$.*

Prime ideals are related to ideals of another important type.

9.6 Definitions. Let L be a lattice and I a proper ideal in L. Then I is said to be a **maximal ideal** if the only ideal properly containing I is L. In other words, I is a maximal ideal if and only if it is a maximal element in $\langle \mathcal{I}(L) \smallsetminus \{L\}; \subseteq \rangle$. A **maximal filter**, more usually known as an **ultrafilter**, is defined dually.

9.7 Theorem. *Let L be a distributive lattice with 1. Then every maximal ideal in L is prime. Dually, in a distributive lattice with 0, every ultrafilter is a prime filter.*

Proof. Let I be a maximal ideal in L and let $a, b \in L$. Assume $a \wedge b \in I$ and $a \notin I$; we require $b \in I$. Define $I_a = \downarrow \{ a \vee c \mid c \in I \}$. Then I_a is an ideal containing I and a (Exercise 9.5). Because I is maximal, we have $I_a = L$. In particular $1 \in I_a$, so $1 = a \vee d$ for some $d \in I$. Then

$$I \ni (a \wedge b) \vee d = (a \vee d) \wedge (b \vee d) = b \vee d.$$

Since $b \leqslant b \vee d$ we have $b \in I$. ∎

In a Boolean lattice we can do better.

9.8 Theorem. *Let B be a Boolean lattice and let I be a proper ideal in B. Then the following are equivalent:*

(i) *I is a maximal ideal;*

(ii) *I is a prime ideal;*

(iii) *for all $a \in B$, it is the case that $a \in I$ if and only if $a' \notin I$.*

Proof. Theorem 9.7 gives (i) $\Rightarrow$ (ii). To prove (ii) $\Rightarrow$ (iii), note that, for any $a \in B$, we have $a \wedge a' = 0$. Because I is prime, $a \in I$ or $a' \in I$. If both a and a' belong to I then $1 = a \vee a' \in I$, ⨏.

Finally we prove that (iii) $\Rightarrow$ (i). Let J be an ideal properly containing I. Fix $a \in J \smallsetminus I$. Then $a' \in I \subseteq J$, so $1 = a \vee a' \in J$. Therefore $J = B$, which shows that I is maximal. ∎

9.9 Ultrafilters on a set. Let S be a non-empty set. An ultrafilter of the Boolean lattice $\wp(S)$ is called an **ultrafilter on** S. Such ultrafilters are important in logic.

An ultrafilter on S is said to be **principal** if it is a principal filter, and **non-principal** otherwise. For each $s \in S$, the set $\{ A \in \wp(S) \mid s \in A \}$ is a principal ultrafilter on S, and every principal ultrafilter is of this form. Non-principal ultrafilters prove much more elusive; see Exercise 9.9. All ultrafilters on a finite set are, of course, principal.

Theorem 9.10 characterizes the ultrafilters on a set. It takes the dual of Theorem 9.8 and adds two other equivalences which are worth stating in this context. For the proof, show (iii) $\Rightarrow$ (v) $\Rightarrow$ (iv) $\Rightarrow$ (i).

9.10 Theorem. *Let $\mathcal{F}$ be a proper filter in $\wp(S)$. Then the following are equivalent:*

(i) *$\mathcal{F}$ is an ultrafilter;*

(ii) *$\mathcal{F}$ is a prime filter;*

(iii) *for each $A \subseteq S$, either $A \in \mathcal{F}$ or $S \setminus A \in \mathcal{F}$;*

(iv) *a subset B of S belongs to $\mathcal{F}$ if $A \cap B \neq \varnothing$ for all $A \in \mathcal{F}$;*

(v) *given pairwise disjoint sets $A_1, \ldots, A_n$ such that $A_1 \cup \cdots \cup A_n = S$, there exists a unique j such that $A_j \in \mathcal{F}$.*

The existence of prime ideals, maximal ideals and ultrafilters

The preceding section is singularly lacking in non-trivial examples. This is unfortunate since in order to remove the finiteness condition in Corollary 9.5 we need a plentiful supply of prime ideals.

To help see what is at stake, consider a Boolean lattice B. Theorem 9.8 implies that a prime ideal in B is just a maximal element of $\langle \mathcal{I}(B) \setminus \{B\}; \subseteq \rangle$. The question of the existence of maximal elements was addressed in Chapter 4. The discussion there shows that this topic has closer affinities with set theory than with lattice theory, and we should not wish knowledge of it to be a prerequisite for proceeding to Chapter 10. The solution to this apparent dilemma is to present a treatment that operates on two levels. The statements (BPI) and (DPI) introduced below assert the existence of certain prime ideals. On one level, (BPI) and (DPI) may be taken as axioms, whose lattice-theoretic implications we pursue. At a deeper level, we show (in a self-contained account which may be omitted) how (BPI) and (DPI) may be derived from (ZL) (4.21). Our remarks in 9.14 will reveal that the difference between these two philosophies is less than might appear.

9.11 (DPI) and (BPI). We consider the following assertions, which embody the existence statements we shall require.

(DPI) Given a distributive lattice L and an ideal J and
a filter G in L such that $J \cap G = \varnothing$, there exist
$I \in \mathcal{I}_p(L)$ and $F = L \smallsetminus I \in \mathcal{F}_p(L)$ such that $J \subseteq I$
and $G \subseteq F$.

(BPI) Given a proper ideal J in a Boolean lattice B, there
exists $I \in \mathcal{I}_p(B)$ such that $J \subseteq I$.

The remainder of this section assumes familiarity with (ZL). The
proof of the first result is typical of Zorn's Lemma arguments (see 4.23).

9.12 Theorem.   (ZL) *implies* (BPI).

Proof. Let B be a Boolean lattice and J be a proper ideal in B. We
apply the special case (ZL)$'$ of (ZL) stated in 4.21 to the set
$$\mathcal{E} := \{\, K \in \mathcal{I}(B) \mid B \ne K \supseteq J \,\},$$
ordered by inclusion. The set $\mathcal{E}$ contains J, and so is non-empty. Let
$\mathcal{C} = \{\, K_\lambda \mid \lambda \in \Lambda \,\}$ be a chain in $\mathcal{E}$. We require $K := \bigcup_{\lambda \in \Lambda} K_\lambda \in \mathcal{E}$.
Certainly $K \ne B$ (why?), $K \supseteq J$ and K is a down-set. It remains
to prove that $a, b \in K$ implies $a \vee b \in K$. For some $\lambda, \mu \in \Lambda$, we
have $a \in K_\lambda$ and $b \in K_\mu$. Since $\mathcal{C}$ is a chain, we may assume without
loss of generality that $K_\lambda \subseteq K_\mu$. But then a, b both belong to K_μ, so
$a \vee b \in K_\mu \subseteq K$. The maximal element of $\mathcal{E}$ provided by (ZL)$'$ is just
the maximal ideal we require (by 9.8).   ∎

The corresponding result for distributive lattices, often referred to
as the **Prime Ideal Theorem**, is slightly more complicated to prove,
but essentially just combines the techniques of 9.7 and 9.12.

9.13 Theorem.   (ZL) *implies* (DPI).

Proof. We take L, G and J as in the statement (DPI) and define
$$\mathcal{E} = \{\, K \in \mathcal{I}(L) \mid K \supseteq J \text{ and } K \cap G = \varnothing \,\}.$$
An argument mildly more complicated than the one in 9.12 shows that
$\langle \mathcal{E}; \subseteq \rangle$ has a maximal element I. It remains to prove that I is prime. To
do this we adapt the proof of 9.7, which is the case $G = \{1\}$. Suppose
$a, b \in L \smallsetminus I$ but $a \wedge b \in I$. Because I is maximal, any ideal properly
containing I is not in $\mathcal{E}$. Consequently $I_a = {\downarrow}\{\, a \vee c \mid c \in I \,\}$ (the
smallest ideal containing I and a) intersects G. Therefore there exists
$c_a \in I$ such that $a \vee c_a$ is above an element of G and hence is itself
in G, because G is an up-set. Similarly we can find $c_b \in I$ such that
$b \vee c_b \in G$. Now consider
$$(a \wedge b) \vee (c_a \vee c_b) = ((a \vee c_a) \vee c_b) \wedge ((b \vee c_b) \vee c_a).$$
The right-hand side is in G, since G is a filter, while the left is in I,
since I is an ideal. This contradicts $I \cap G = \varnothing$. ↯.   ∎

9.14 Remarks. Some further comments on the relationship between (ZL), (BPI) and (DPI) are appropriate, although we cannot attempt to justify all the assertions we make. Consult [18] and [25] for more details, related results and references.

When L is a distributive lattice with 1, we may take $G = \{1\}$ in (DPI). Then (DPI) implies the existence of a maximal ideal of L containing a given proper ideal J. So (DPI), restricted to Boolean lattices, yields (BPI) as a special case. Much less obviously, (BPI) $\Rightarrow$ (DPI). This is proved by constructing an embedding of a given distributive lattice into a Boolean lattice, to which (BPI) is applied. Hence (BPI) and (DPI) are equivalent.

We proved in Chapter 4 that (ZL) is equivalent to the Axiom of Choice, (AC), and remarked that many other equivalents of (AC) (set-theoretic and otherwise) were known. It turns out that one such statement is (DMI):

(DMI) Every distributive lattice with 1 contains a maximal
       ideal.

It is easy to derive (DMI) from (ZL) (see Exercise 4.19). Conversely it can be proved that (AC) can be derived from (DMI), applied to a suitable lattice of sets. A proof that (DMI) $\Rightarrow$ (DPI) is indicated in Exercise 9.11.

By contrast, (BPI) and (DPI) belong to a family of conditions known to be equivalent to the choice principle $(AC)_F$ (asserting that every family of non-empty *finite* sets has a choice function). It is known that $(AC)_F$ is stricly weaker than (AC), so that it is not true that (DPI) $\Rightarrow$ (DMI). However $(AC)_F$ is not derivable within traditional Zermelo–Fraenkel set theory. To obtain results such as (DPI) and (BPI) some additional axiom must be added, and whether this is (AC), (ZL), or even (DPI) itself, is a matter of choice. Thus our suggestion that readers ignorant of (ZL) should take (DPI) as a hypothesis had a sound logical basis.

There are many instances in logic and topology of the construction of ultrafilters in Boolean lattices (especially powerset lattices). The proof via filters of Tychonoff's Theorem is an example. Accordingly we introduce

(BUF) Given a proper filter G in a Boolean lattice B, there
       exists $F \in \mathcal{F}_p(B)$ such that $G \subseteq F$.

A proper filter (an ultrafilter) in a Boolean lattice B is a proper ideal (a maximal ideal) in B^{∂} (which is also a Boolean lattice). Thus the statements (BPI) and (BUF) are equivalent.

We sum up the connections between the various conditions below.

$$\text{(AC)} \iff \text{(ZL)} \implies \text{(BPI)} \iff \text{(BUF)}$$

$$\Updownarrow \qquad\qquad\qquad \Updownarrow \qquad\qquad \Updownarrow$$

$$\text{(DMI)} \qquad\qquad \implies \qquad \text{(BMI)} \iff \text{(AC)}_F$$

9.15 Postscript: LINDA again. We promised in Chapter 7 that we would prove the Adequacy Theorem for the system **L** of propositional calculus by using the Boolean algebra $LA_\vdash$ (recall 7.18).

We wish to prove that $\vdash_\mathbf{L} \varphi$ for every tautology φ. We prove the contrapositive. Suppose $\nvdash_\mathbf{L} \varphi$. Then $[\varphi] \neq 1$ in $LA_\vdash$. The ideal $\downarrow[\varphi]$ is proper. We appeal to (BPI) to find a prime ideal I such that $\downarrow[\varphi] \subseteq I$. Define a map from the wffs to $\{\mathbf{F}, \mathbf{T}\}$ by

$$v(\psi) = \begin{cases} \mathbf{T} & \text{if } [\psi] \notin I, \\ \mathbf{F} & \text{if } [\psi] \in I. \end{cases}$$

It can be proved directly that this map is a valuation (see 7.17); alternatively this may be verified by introducing as a stepping stone the Boolean homomorphism f on $LA_\vdash$ specified by $f^{-1}(0) = I$. By construction, $v(\varphi) = \mathbf{F}$, so φ is not a tautology. This completes the proof of the Adequacy Theorem.

It is usual in elementary treatments of propositional calculus to take the set of propositional variables to be countable. This assumption is made to avoid having to invoke (ZL). When the set of propositional variables is countable, so is $LA_\vdash$. Exercise 9.8 shows that (BPI) for countable Boolean algebras can be proved without (ZL) (by a process remarkably reminiscent of the technique customarily employed in the proof of the Adequacy Theorem).

Exercises

Exercises from the text. Prove the assertions in 9.10. Fill in the (ZL) part of the proof of Theorem 9.13.

9.1 Let L be a lattice. Prove that $\mathcal{I}_p(L) \cong \mathcal{F}_p(L)^\partial$.

9.2 Let L and K be bounded lattices and $f \colon L \to K$ a $\{0,1\}$-homomorphism.
   (i) Show that $f^{-1}(0)$ is an ideal in L.
   (ii) Show that, if $K = \mathbf{2}$, then $f^{-1}(0)$ is a prime ideal in L.

(iii) Let I be a prime ideal in L. Define $f_I \colon L \to \mathbf{2}$ by

$$f_I(a) = \begin{cases} 1 & \text{if } a \notin I, \\ 0 & \text{if } a \in I. \end{cases}$$

Prove that f_I is a $\{0,1\}$-homomorphism.

(iv) Let X denote the set of all $\{0,1\}$-homomorphisms from L to $\mathbf{2}$, ordered pointwise. Set up an order-isomorphism between $\mathcal{I}_p(L)$, ordered by inclusion, and X^∂. [cf. Exercise 8.19 and Corollary 9.5.]

9.3 Let L and K be bounded distributive lattices.

  (i) Prove that every ideal of $L \times K$ is of the form $I \times J$ where I is an ideal of L and J is an ideal of K.

  (ii) Let I be a prime ideal of L and let J be a prime ideal of K. Show that $I \times K$ and $L \times J$ are prime ideals in $L \times K$ and that every prime ideal of $L \times K$ is of this form.

9.4 Let B be a Boolean algebra. Prove that I is a lattice ideal in B if and only if I is a ring ideal in the ring $\langle B; +, \cdot \rangle$ defined in Exercise 7.8.

9.5 Let I be an ideal in a lattice L and let $a \in L$. Show that the set $\downarrow\{a \vee c \mid c \in I\}$ is an ideal and is the smallest ideal in L containing both I and a.

9.6 Find all ideals in (i) $\mathbb{Z}$, (ii) $\mathbb{Q}$ (with their usual orders). Which of these ideals are principal?

9.7 Find all prime ideals (prime filters) in $\langle \mathbb{N}_0; \mathrm{lcm}, \mathrm{gcd} \rangle$. [See Proposition 9.4 and Exercise 8.14.] Describe the order on the set of prime filters of $\langle \mathbb{N}_0; \mathrm{lcm}, \mathrm{gcd} \rangle$.

9.8 Let $B = \{b_1, b_2, \dots, b_n, \dots\}$ be a countable Boolean lattice. Without using (ZL) or an equivalent, prove that B satisfies (BUF). [Hint. Consider $\bigcup_{n \geqslant 1}(\{b_1, b_2, \dots, b_n\}].$]

9.9 Let S be an infinite set.

  (i) Let $\mathcal{G}$ be the set of cofinite subsets of S. (Recall that $A \subseteq S$ is called **cofinite** if $S \setminus A$ is finite.)

    (a) Show that $\mathcal{G}$ is a filter in $\wp(S)$.

    (b) Show that if $\mathcal{F}$ is a proper filter in $\wp(S)$ and $\mathcal{G} \subseteq \mathcal{F}$, then $\mathcal{F}$ is not principal.

(ii) Assume that (BUF) holds. Show that there is a non-principal ultrafilter on S.

(iii) Assume that (ZL) holds. Prove directly from (ZL), or (ZL)$'$, that there is a non-principal ultrafilter on S.

9.10 A filter G of a lattice L is called **distributive** if it satisfies

$$(\forall a, b, c \in L)\, a \vee b, a \vee c \in G \Longrightarrow a \vee (b \wedge c) \in G.$$

(i) Find all distributive filters of $\mathbf{N}_5$ and $\mathbf{M}_5$.

(ii) Prove that L is distributive if and only if every filter of L is distributive.

(iii) Let L be a lattice and G a filter in L. Prove that the following are equivalent:

    (a) G is a distributive filter;

    (b) every ideal I which is a maximal element of $\{\, K \in \mathcal{I}(L) \mid K \cap G = \varnothing \,\}$ is a prime ideal;

    (c) G is an intersection of prime filters, that is, $G = \bigcap_{i \in I} F_i$ for some family $\{F_i\}_{i \in I}$ of prime filters.

[Hint. The implication (a) $\Rightarrow$ (b) is a refinement of the second portion of the proof of Theorem 9.13, while (b) $\Rightarrow$ (c) is an easy consequence of Exercise 9.10(i) (Note that (ZL) is required.)]

9.11 Let L be a distributive lattice, J an ideal and G a filter of L such that $J \cap G = \varnothing$.

(i) Suppose that there is an onto homomorphism $f \colon L \to K$ such that

    (a) $|K| \geqslant 2$ and K has a 0 and a 1;

    (b) $J \subseteq f^{-1}(\{0\})$ and $G \subseteq f^{-1}(\{1\})$.

Show that (DMI) applied to K yields (DPI) for L.

(ii) Let θ_J be the congruence on L defined in Exercise 6.19 and let θ^G be defined dually. Show that the quotient map $f = q \colon L \to L/(\theta_J \vee \theta^G) = K$ satisfies (a) and (b) of (i).

Thus (DMI) implies (DPI).

10

Representation Theory: the General Case

This chapter contains representation theorems for arbitrary Boolean algebras and bounded distributive lattices. Setting up the theory requires more sophisticated mathematics than in the finite case. Once obtained, however, the representation theorems are often as simple to use as their finite versions. We go on to develop the associated duality and present a sample of elementary applications. The exercises give further indication of the potential of duality theory.

Representation by lattices of sets

The following lemma is a slight refinement of Proposition 9.4. It gives the clue on how to extend the representation theorems in Chapter 8 and will show those results to be special cases of those we shortly obtain.

10.1 Lemma.

(i) Let L be a finite distributive lattice and let $a \in L$. Then the map $x \mapsto L \setminus {\uparrow} x$ is an order-isomorphism of $\{ x \in \mathcal{J}(L) \mid x \leqslant a \}$ onto $\{ I \in \mathcal{I}_p(L) \mid a \notin I \}$.

(ii) Let B be a finite Boolean algebra and let $a \in B$. Then the map $x \mapsto B \setminus {\uparrow} x$ is a bijection of $\{ x \in \mathcal{A}(L) \mid x \leqslant a \}$ onto $\{ I \in \mathcal{I}_p(B) \mid a \notin I \}$.

10.2 Lemma. Let L be a lattice and let $X = \mathcal{I}_p(L)$. Then the map $\eta : L \to \wp(X)$ defined by

$$\eta : a \longmapsto X_a := \{ I \in \mathcal{I}_p(L) \mid a \notin I \}$$

is a lattice homomorphism.

Proof. We have to show that $X_{a \vee b} = X_a \cup X_b$ and $X_{a \wedge b} = X_a \cap X_b$, for all $a, b \in L$. Take $I \in \mathcal{I}_p(L)$. Since I is an ideal,

$$a \vee b \in I \text{ if and only if } a \in I \text{ and } b \in I$$

and, since I is prime,

$$a \wedge b \in I \text{ if and only if } a \in I \text{ or } b \in I.$$

Thus we have

$$X_{a \vee b} = \{ I \in \mathcal{I}_p(L) \mid a \vee b \notin I \}$$
$$= \{ I \in \mathcal{I}_p(L) \mid a \notin I \text{ or } b \notin I \}$$
$$= X_a \cup X_b.$$

Similarly, $X_{a \wedge b} = X_a \cap X_b$.   ∎

We should like the map η to give a faithful copy of L in the lattice $\wp(\mathcal{I}_p(L))$. We certainly cannot prove this without the additional hypothesis of distributivity, because a lattice of sets must be distributive. Theorem 10.3 shows that (DPI) is exactly what is needed to ensure that a distributive lattice L has enough prime ideals for $\eta \colon L \to \wp(\mathcal{I}_p(L))$ to be an embedding.

10.3 Theorem.   *Let L be a lattice. Then the following are equivalent:*

(i) *L is distributive;*

(ii) *given an ideal J of L and a filter G of L with $J \cap G = \varnothing$, there exists a prime ideal I such that $J \subseteq I$ and $I \cap G = \varnothing$;*

(iii) *given $a, b \in L$ with $a \not\leqslant b$, there exists a prime ideal I such that $a \notin I$ and $b \in I$;*

(iv) *the map $\eta \colon a \mapsto X_a := \{ I \in \mathcal{I}_p(L) \mid a \notin I \}$ is an embedding of L into $\wp(\mathcal{I}_p(L))$;*

(v) *L is isomorphic to a lattice of sets.*

Proof. The implications (iv) $\Rightarrow$ (v) $\Rightarrow$ (i) are trivial and (i) $\Rightarrow$ (ii) is the statement that (DPI) holds. To prove (ii) $\Rightarrow$ (iii) just take $J = {\downarrow}a$ and $G = {\uparrow}b$ in (DPI).

Since η is a homomorphism, it is order-preserving. To prove that (iii) $\Rightarrow$ (iv) it is enough to show that $a \not\leqslant b$ implies $X_a \not\subseteq X_b$. This is true since the prime ideal I supplied by (iii) belongs to $X_a \setminus X_b$.   ∎

For Boolean algebras we have the following result.

10.4 Theorem.   *Let B be a Boolean algebra. Then*

(i) *given a proper ideal J in B there exists a maximal ideal $I \in \mathcal{I}_p(B)$ with $J \subseteq I$;*

(ii) *given $a \neq b$ in B, there exists a maximal ideal $I \in \mathcal{I}_p(B)$ such that I contains one and only one of a and b;*

(iii) *there is a Boolean algebra isomorphism from B onto a subalgebra of $\wp(X)$, where $X = \mathcal{I}_p(B)$.*

Proof. We first show that (i), which is the statement (BPI) for B, implies (ii). Take $a, b \in B$ with $a \neq b$. Without loss of generality we may assume $a \not\leqslant b$ and this gives $1 \neq a' \vee b$. Apply (i) with $J = {\downarrow}(a' \vee b)$. Any prime ideal I containing J contains b, but, by Theorem 9.8, not a.

The map $\eta \colon a \mapsto X_a := \{ I \in I_p(B) \mid a \notin I \}$ is a lattice homomorphism. Also $X_0 = \varnothing$ because each prime ideal contains 0 and $X_1 = X$

since each prime ideal is proper. So, by 7.5, η is a Boolean algebra homomorphism. If (ii) holds it is also one-to-one. ∎

The prime ideal space

We have set out to prove analogues of Theorems 8.3 and 8.17 not restricted to finite lattices. We have in part achieved this by establishing that any distributive lattice is isomorphic to a lattice of sets and any Boolean algebra to an algebra of sets. In the finite case we were able to be more explicit. What Theorems 10.3 and 10.4 lack is a characterization of the image of the embedding η. We can however say straightaway that η cannot always be onto. We noted in 7.7(2) that not every Boolean algebra is isomorphic to a powerset algebra and a similar argument shows that not every distributive lattice is isomorphic to a lattice $\mathcal{O}(P)$. The description of $\operatorname{im}\eta$ has to be in terms of additional structure on the set of prime ideals. A topological structure is exactly what we need. A **topology** on a set X is a family of subsets of X containing X and $\varnothing$ and closed under arbitrary unions and finite intersections. Readers whose knowledge of topology is rusty or non-existent will find an outline of the concepts and results we need in the appendix to this chapter. References such as 10.1A are to this appendix.

The Boolean case is somewhat simpler than the distributive lattice case, so we consider it first.

10.5 Topologizing the prime ideals of a Boolean lattice. The family of clopen subsets of a topological space $\langle X; \mathcal{T} \rangle$ forms a Boolean lattice. This suggests that, given a Boolean lattice B, we might try to impose a topology, $\mathcal{T}$, on $\mathcal{I}_p(B)$ so that $\operatorname{im}\eta$ is characterized as the family of clopen subsets of $\langle \mathcal{I}_p(B); \mathcal{T} \rangle$. It is certainly necessary that

$$X_a := \{\, I \in \mathcal{I}_p(B) \mid a \notin I \,\}$$

be in $\mathcal{T}$ for each $a \in B$. The family $\mathcal{B} := \{\, X_a \mid a \in B \,\}$ is not a topology because it is not closed under the formation of arbitrary unions. We have to define $\mathcal{T}$ on $X := \mathcal{I}_p(B)$ as follows:

$$\mathcal{T} := \{\, U \subseteq X \mid U \text{ is a union of members of } \mathcal{B} \,\}.$$

In the terminology of 10.3A, $\mathcal{B}$ is a basis for $\mathcal{T}$ (which is indeed a topology). The topological space $\langle \mathcal{I}_p(B); \mathcal{T} \rangle$ is called the **prime ideal space** or **dual space** of B.

Each element of $\mathcal{B}$ is clopen in X, because $X \smallsetminus X_a = X_{a'}$ and so $X \smallsetminus X_a$ is open. To prove that *every* clopen subset of $\langle X; \mathcal{T} \rangle$ is of the form X_a, we need further information about the prime ideal space.

10.6 Proposition. *Let B be a Boolean lattice and let $X := \mathcal{I}_p(B)$, where $\langle \mathcal{I}_p(B); \mathcal{T} \rangle$ is the prime ideal space of B. Then X is compact.*

Proof. Let $\mathcal{U}$ be an open cover of X. We have to show that there exist finitely many members of $\mathcal{U}$ whose union is X. Every open set is a union of sets X_a and we may therefore assume without loss of generality that $\mathcal{U} \subseteq B$. Write $\mathcal{U} = \{ X_a \mid a \in A \}$, where $A \subseteq B$. Let J be the smallest ideal containing A, that is (by Exercise 2.13),

$$J = \{ b \in B \mid b \leqslant a_1 \vee \ldots \vee a_n \text{ for some } a_1, \ldots, a_n \in A \}.$$

If J is not proper, then $1 \in J$ and we have $a_1 \vee \ldots \vee a_n = 1$ for some finite subset $\{a_1, \ldots, a_n\}$ of A. Then $X = X_1 = X_{a_1 \vee \ldots \vee a_n} = X_{a_1} \cup \cdots \cup X_{a_n}$ and $\{X_{a_i}\}_{1 \leqslant i \leqslant n}$ provides the required finite subcover of $\mathcal{U}$.

If J is proper we can use (BPI) to obtain a prime ideal I containing J. But then I belongs to X but to no member of $\mathcal{U}$, ⚡. ∎

10.7 Proposition. *Let $\langle X; \mathcal{T} \rangle$ be the prime ideal space of the Boolean lattice B. Then the clopen subsets of X are exactly the sets X_a for $a \in B$. Further, given distinct points $x, y \in X$, there exists a clopen subset V of X such that $x \in V$ and $y \notin V$.*

Proof. We have already noted that each set X_a is clopen. Also, given distinct elements I_1 and I_2 of $\mathcal{I}_p(B)$, there exists, without loss of generality, $a \in I_1 \smallsetminus I_2$. Then X_a contains I_2 but not I_1. This proves the final assertion.

It remains to prove that an arbitrary clopen subset, U, of X is of the form X_a for some $a \in B$. Because U is open, $U = \bigcup_{a \in A} X_a$ for some subset A of B. But U is also a closed subset of X and so compact (by 10.7A). Hence there exists a finite subset A_1 of A such that $U = \bigcup_{a \in A_1} X_a$. Then $U = X_a$, where $a = \bigvee A_1$ (see 10.2). ∎

By combining Theorem 10.3 and the first part of the preceding lemma we obtain Stone's famous representation theorem.

10.8 Stone's representation theorem for Boolean algebras. *Let B be a Boolean algebra. Then the map*

$$\eta \colon a \longmapsto X_a := \{ I \in \mathcal{I}_p(B) \mid a \notin I \}$$

is a Boolean algebra isomorphism of B onto the Boolean algebra of clopen subsets of the dual space $\langle \mathcal{I}_p(B); \mathcal{T} \rangle$ of B.

To exploit this representation to the full we need to know more about topological spaces with the properties possessed by $\mathcal{I}_p(B)$. The last part of Proposition 10.7 asserts that the prime ideal space of a Boolean lattice satisfies a separation condition guaranteeing that the space has 'plenty' of clopen subsets. We next pursue the topological ramifications of this condition.

10.9 Totally disconnected spaces and Boolean spaces. We say that a topological space $\langle X; \mathcal{T} \rangle$ is **totally disconnected** if, given distinct points $x, y \in X$, there exists a clopen subset V of X such that $x \in V$ and $y \notin V$. Topologists usually give a different definition of total disconnectedness. For compact spaces their definition agrees with ours. If $\langle X; \mathcal{T} \rangle$ is both compact and totally disconnected it is said to be a **Boolean space**. Propositions 10.6 and 10.7 assert that $\langle \mathcal{I}_p(B); \mathcal{T} \rangle$ is a Boolean space for every Boolean lattice B. We denote by $\mathcal{P}(X)$ the family of clopen subsets of a Boolean space X.

Given distinct points x, y in a totally disconnected space X, there exist disjoint clopen sets V and $W := X \setminus V$ such that $x \in V$ and $y \in W$. This implies that a totally disconnected space is Hausdorff; see 10.5A. In particular a Boolean space is compact and Hausdorff. Compact Hausdorff spaces are well known to have many nice properties, some of which are stated in 10.7A and 10.8A. Working with Boolean spaces is like working with compact Hausdorff spaces, but with the bonus of having a basis of clopen sets. This is illustrated by the proof of Lemma 10.10 which is analogous to that of Lemma 10.8A, except that use of the total disconnectedness condition replaces use of the Hausdorff condition.

10.10 Lemma. *Let $\langle X; \mathcal{T} \rangle$ be a Boolean space.*

(i) *Let Y be a closed subset of X and $x \notin Y$. Then there exists a clopen set V such that $Y \subseteq V$ and $x \notin V$.*

(ii) *Let Y and Z be disjoint closed subsets of X. Then there exists a clopen set U such that $Y \subseteq U$ and $Z \cap U = \varnothing$.*

Proof. (i) For each $y \in Y$ there exists a clopen set V_y such that $y \in V_y$ and $x \notin V_y$. The open sets $\{ V_y \mid y \in Y \}$ form an open cover of Y. Since Y is compact (by 10.7A) there exist $y_1, \dots, y_n$ such that $Y \subseteq V := V_{y_1} \cup \cdots \cup V_{y_n}$. As a finite union of clopen sets, V is clopen; by construction it does not contain x.

(ii) This is left as an exercise; compare Lemma 10.8A. ∎

This section has so far been totally bereft of examples. The simplest example of a Boolean space is obtained by taking a finite set with the discrete topology. A less trivial example is given below.

10.11 Example. Denote by $\mathbb{N}_\infty$ the set of natural numbers with an additional point, ∞, adjoined. We define $\mathcal{T}$ as follows: a subset U of $\mathbb{N}_\infty$ belongs to $\mathcal{T}$ if

$$\text{either (a) } \infty \notin U,$$
$$\text{or (b) } \infty \in U \text{ and } \mathbb{N}_\infty \setminus U \text{ is finite.}$$

We leave as an exercise the proof that $\mathcal{T}$ is a topology. A subset V of $\mathbb{N}_\infty$ is clopen if and only if V and $\mathbb{N}_\infty \setminus V$ both satisfy conditions (a) and (b). It follows that the clopen subsets of $\mathbb{N}_\infty$ are the finite sets not containing ∞ and their complements.

It is now easy to show that $\mathbb{N}_\infty$ is totally disconnected. Given distinct points $x, y \in \mathbb{N}_\infty$, we may assume without loss of generality that $x \neq \infty$. Then $\{x\}$ is clopen and contains x but not y.

We next prove that $\mathbb{N}_\infty$ is compact. Take an open cover $\mathcal{U}$ of $\mathbb{N}_\infty$. Some member of $\mathcal{U}$ must contain ∞; say U is such a set. Then $\mathbb{N}_\infty \setminus U$ is finite, by (b). Hence only finitely many members of $\mathcal{U}$ are needed to cover $\mathbb{N}_\infty \setminus U$ and these, together with U, provide the required finite subcover of $\mathcal{U}$. [The *cognoscenti* will recognize the space $\mathbb{N}_\infty$ as the 1-point compactifcation of a countable discrete space.]

We later discover that the Boolean space $\mathbb{N}_\infty$ is (homeomorphic to) the prime ideal space of the finite-cofinite algebra $\mathrm{FC}(\mathbb{N})$ (see 7.7(2)).

We now return to distributive lattices, which we temporarily abandoned at the beginning of this section.

10.12 Stocktaking. Let L be a distributive lattice and let $X = \mathcal{I}_p(L)$ be its set of prime ideals, ordered as usual by inclusion. We already have representations for L in two special cases.

When L is Boolean and X is topologized in the way described above, L is isomorphic to the algebra $\mathcal{P}(X)$ of clopen subsets of X. Every prime ideal of a Boolean lattice is maximal (by 9.8), so the order on X is discrete (that is, $x \leqslant y$ in X if and only if $x = y$). Thus the order has no active role in this case.

When L is finite, L is isomorphic to the lattice $\mathcal{O}(X)$ of down-sets of X (by 8.17 and 9.4). Suppose X has a topology, $\mathcal{T}$, making it a Boolean space. Then $\mathcal{T}$ is the discrete topology, in which every subset is clopen (see 10.9A). In this case the topology contributes nothing.

These observations suggest that to represent L in general we should equip X with the inclusion order *and* a suitable Boolean space topology. A prime candidate for a lattice isomorphic to L would then be the lattice of all clopen down-sets of X. Our remarks above imply that this lattice coincides with $\mathcal{O}(X)$ when L is finite, with $\mathcal{P}(X)$ when L is Boolean (and with $\wp(X)$ when L is both finite and Boolean). We prove in 10.18 that a bounded distributive lattice L is indeed isomorphic to the lattice of clopen down-sets of $\mathcal{I}_p(L)$, ordered by inclusion and appropriately topologized, and thereby obtain a natural common generalization of Birkhoff's and Stone's theorems. The boundedness restriction is forced upon us because the lattice of clopen down-sets is bounded. Extensions of the theorem to lattices lacking bounds do exist, but we do not consider them here.

10.13 The prime ideal space of a bounded distributive lattice. Let L be a distributive lattice with 0 and 1 and for each $a \in L$ let

$$X_a := \{ I \in \mathcal{I}_p(L) \mid a \notin I \},$$

as before. Let $X := \mathcal{I}_p(L)$. We want a topology, $\mathcal{T}$, on X so that each X_a is clopen. Accordingly, we want every element of

$$\mathcal{S} := \{ X_b \mid b \in L \} \cup \{ X \smallsetminus X_c \mid c \in L \}$$

to be in $\mathcal{T}$. Compared with the Boolean case we have a double complication to contend with. The family $\mathcal{S}$ contains sets of two types and it is also not closed under finite intersections. We let $\mathcal{B} := \{ X_b \cap (X \smallsetminus X_c) \mid b, c \in L \}$. Since L has 0 and 1, $\mathcal{B}$ contains $\mathcal{S}$. Also $\mathcal{B}$ is closed under finite intersections. Finally we define $\mathcal{T}$ as follows: $U \in \mathcal{T}$ if U is a union of members of $\mathcal{B}$. Then $\mathcal{T}$ is the smallest topology containing $\mathcal{S}$; in the terminology of 10.3A, $\mathcal{S}$ is a subbasis for $\mathcal{T}$ and $\mathcal{B}$ a basis.

10.14 Theorem. *Let L be a bounded distributive lattice. Then the prime ideal space $\langle \mathcal{I}_p(L); \mathcal{T} \rangle$ is compact.*

Proof. Alexander's Subbasis Lemma proved (with the aid of (BPI)) in 10.10A tells us that it is sufficient to prove that any open cover $\mathcal{U}$ of $X = \mathcal{I}_p(L)$ by sets in the subbasis $\mathcal{S}$ has a finite subcover. Doing this is only slightly more complicated than proving 10.6. Let

$$\mathcal{U} = \{ X_b \mid b \in A_0 \} \cup \{ X \smallsetminus X_c \mid c \in A_1 \}.$$

Let J be the ideal generated by A_0 (this is $\{0\}$ if A_0 is empty) and let G be the filter generated by A_1 (this is $\{1\}$ if A_1 is empty). Assume first that $J \cap G = \varnothing$ and invoke (DPI) to find a prime ideal I such that $J \subseteq I$ and $G \cap I = \varnothing$. Then $I \notin X_b$ for any $b \in A_0$ and $I \notin X \smallsetminus X_c$ for any $c \in A_1$ and this means that $\mathcal{U}$ does not cover X, $\lightning$.

Hence $J \cap G \neq \varnothing$. Take $a \in J \cap G$. If A_0 and A_1 are both non-empty there exist $b_1, \ldots, b_j \in A_0$ and $c_1, \ldots, c_k \in A_1$ such that

$$c_1 \wedge \ldots \wedge c_k \leqslant a \leqslant b_1 \vee \ldots \vee b_j,$$

whence

$$X = X_1 = X_{b_1} \cup \cdots \cup X_{b_j} \cup (X \smallsetminus X_{c_1}) \cup \cdots \cup (X \smallsetminus X_{c_k}).$$

Thus in this case $\mathcal{U}$ has a finite subcover. The case $A_1 = \varnothing$ is treated as in 10.6 and the case $A_0 = \varnothing$ is similar. ∎

10.15 Totally order-disconnected spaces. A set X carrying a topology, $\mathcal{T}$, and an order relation, $\leqslant$, is called an **ordered (topological) space** and denoted $\langle X; \mathcal{T}, \leqslant \rangle$ (or by X where no ambiguity would result). It is said to be **totally order-disconnected** if, given $x, y \in X$ with $x \not\geqslant y$, there exists a clopen down-set U such that $x \in U$ and $y \notin U$. This separation condition is illustrated in Figure 10.1. We call a compact totally order-disconnected space a **CTOD space**; CTOD spaces are also known as ordered Stone spaces or Priestley spaces. We shall denote by $\mathcal{O}(X)$ the family of clopen down-sets of a CTOD space X. We have previously used $\mathcal{O}(X)$ to mean the family of all down-sets. No conflict arises since we shall not need the latter in connection with CTOD spaces except in the finite case, where the two usages agree (by 10.9A).

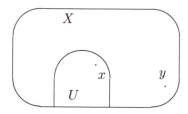

Figure 10.1

Clearly total order-disconnectedness implies total disconnectedness and the two notions coincide when the order is discrete. In many ways CTOD spaces behave like a cross between Boolean spaces and ordered sets. Lemma 10.16 illustrates this. The first part, which is an easy exercise, should be compared with Lemma 1.21. The remainder of the lemma is an analogue for CTOD spaces of Lemma 10.10 and is proved in the same way, but with clopen down-sets replacing clopen sets.

10.16 Lemma. Let $\langle X; \mathcal{T}, \leqslant \rangle$ be a CTOD space.

(i) $x \leqslant y$ in X if and only if $y \in U$ implies $x \in U$ for every $U \in \mathcal{O}(X)$.

(ii) (a) Let Y be a closed down-set in X and let $x \notin Y$. Then there exists a clopen down-set U such that $Y \subseteq U$ and $x \notin U$.

(b) *Let* Y *and* Z *be disjoint closed subsets of* X *such that* Y *is a down-set and* Z *is an up-set. Then there exists a clopen down-set* U *such that* $Y \subseteq U$ *and* $Z \cap U = \emptyset$.

We can now characterize clopen sets and clopen down-sets in the dual space, $\langle \mathcal{I}_p(L); \mathcal{T}, \subseteq \rangle$, of a bounded distributive lattice, L. The proof follows the same lines as that of Proposition 10.7 and is left as an exercise.

10.17 Lemma. *Let* L *be a bounded distributive lattice with dual space* $\langle X; \mathcal{T}, \subseteq \rangle$, *where* $X = \mathcal{I}_p(L)$. *Then*

(i) *the clopen subsets of* X *are the sets* $X_b \cap (X \smallsetminus X_c)$ *for* $b, c \in L$;

(ii) *the clopen down-sets of* X *are exactly the sets* X_a *for* $a \in L$.

10.18 Priestley's representation theorem for distributive lattices. *Let* L *be a bounded distributive lattice. Then the map*

$$\eta \colon a \longmapsto X_a := \{\, I \in \mathcal{I}_p(L) \mid a \notin I \,\}$$

is an isomorphism of L *onto the lattice of clopen down-sets of the dual space* $\langle \mathcal{I}_p(L); \mathcal{T}, \subseteq \rangle$ *of* L.

Proof. Combine Theorem 10.3 and Lemma 10.17(ii). ∎

Duality

The remainder of the chapter parallels the final section of Chapter 8. So far in this chapter we have presented Boolean algebra results first, and then their distributive lattice counterparts. While we were setting up the necessary machinery this had advantages, since the Boolean case was somewhat simpler. Also this approach gave an easily identified shortest-path route to Stone's Theorem for those interested primarily in Boolean algebras. We contend that, once obtained, the distributive lattice representation is just as easy to work with as that for Boolean algebras and also provides a richer source of examples. Therefore, we henceforth place the emphasis on distributive lattices, deriving Boolean algebra results as corollaries by specializing to the discrete order.

Our first task is to generalize Theorem 8.19. The proof uses several topological lemmas. However the benefits we shall reap from the result make worthwhile the hard work involved in the proof. Ordered spaces X and Y are 'essentially the same' if there exists a map φ from X onto Y which is simultaneously an order-isomorphism and a homeomorphism. We call such a map an **order-homeomorphism** and say X and Y are **order-homeomorphic**.

10.19 Theorem.

(i) *Let Y be a Boolean space, let B be the algebra $\mathcal{P}(Y)$ of clopen subsets of Y and let X be the dual space of B. Then Y and X are homeomorphic.*

(ii) *Let Y be a CTOD space, let L be the lattice $\mathcal{O}(Y)$ of clopen down-sets of Y and let X be the dual space of L. Then Y and X are order-homeomorphic.*

Proof. Part (i) is essentially the special case of (ii) in which the order is discrete. We prove (ii). We define $\varepsilon \colon Y \to X$ by

$$\varepsilon(y) := \{\, a \in L \mid y \notin a \,\}.$$

Certainly $\varepsilon(y)$ is a prime ideal in L. We shall show

(a) ε is an order-embedding (and hence, by 1.13(3), one-to-one);

(b) ε is continuous;

(c) ε maps Y onto X.

By 10.7A this will establish (ii).

For (a) note that

$$y \leqslant z \text{ in } Y \iff (\forall a \in L)\,(z \in a \Rightarrow y \in a) \quad \text{(by Lemma 1.16(i))}$$
$$\iff \varepsilon(y) \subseteq \varepsilon(z).$$

To prove (b) we need Lemma 10.4A. It implies that (b) holds so long as $\varepsilon^{-1}(X_a)$ and $\varepsilon^{-1}(X \smallsetminus X_a)$ are open for each $a \in L$. But

$$\varepsilon^{-1}(X \smallsetminus X_a) = \{\, y \in Y \mid \varepsilon(y) \notin X_a \,\} = Y \smallsetminus \varepsilon^{-1}(X_a).$$

Thus (b) holds provided $\varepsilon^{-1}(X_a)$ is clopen in Y for each $a \in L$. But

$$\varepsilon^{-1}(X_a) = \{\, y \in Y \mid \varepsilon(y) \in X_a \,\}$$
$$= \{\, y \in Y \mid a \notin \varepsilon(y) \,\} \qquad \text{(by the definition of } X_a)$$
$$= a \qquad\qquad\qquad \text{(by the definition of } \varepsilon),$$

and this is clopen, by the definition of L.

Finally we prove (c). By Lemma 10.7A, $\varepsilon(Y)$ is a closed subset of X. Suppose by way of contradiction that there exists $x \in X \smallsetminus \varepsilon(Y)$. Lemma 10.16(ii) implies that there is a clopen subset V of X such that $\varepsilon(Y) \subseteq V$ and $x \notin V$. By 10.17, $V = X_b \cap (X \smallsetminus X_c)$ for some $b, c \in L$. From above we have $\varepsilon^{-1}(V) = b \cap (Y \smallsetminus c)$. Since $\varepsilon(Y) \subseteq V$, this implies $Y = b \cap (Y \smallsetminus c)$, from which it follows that $X_b \cap (X \smallsetminus X_c) = \varnothing$, ⚡. ∎

It is far from easy to find all the prime ideals of an infinite lattice and to describe the structure of the dual space. Fortunately the following corollary to Theorem 10.19 provides an indirect way to obtain dual spaces.

10.20 Theorem.

(i) *Let B be a Boolean algebra and Y a Boolean space such that $\mathcal{P}(Y) \cong B$. Then the dual space of B is (homeomorphic to) Y.*

(ii) *Let L be a bounded distributive lattice and Y a CTOD space such that $\mathcal{O}(Y) \cong L$. Then the dual space of L is (order-homeomorphic to) Y.*

Our first dual space examples below involve a minimum of topology. We include Examples (3) and (4) for those with the topological knowledge and sophistication to appreciate them. They extend our list of Boolean spaces, but we shall not use them later.

10.21 Examples.

(1) Consider again the Boolean space $\mathbb{N}_\infty$ introduced in 10.11. Its algebra B of clopen sets consists of the finite sets not containing ∞ and their complements. Define $f \colon \mathrm{FC}(\mathbb{N}) \to B$ by

$$f(a) = \begin{cases} a & \text{if } a \text{ is finite,} \\ a \cup \{\infty\} & \text{if } a \text{ is cofinite.} \end{cases}$$

This map is easily seen to be an isomorphism. Therefore the dual space of $\mathrm{FC}(\mathbb{N})$ can be identified with $\mathbb{N}_\infty$. We can now recognize the elements of $\mathcal{I}_p(B)$. The points of $\mathbb{N}$ are in one-to-one correspondence with the principal prime ideals of $\mathrm{FC}(\mathbb{N})$, via the map $n \mapsto \downarrow(\mathbb{N} \setminus \{n\})$. There is a single non-principal prime ideal, associated with ∞: it consists of all finite subsets of $\mathbb{N}$.

(2) A variety of CTOD spaces can be obtained by equipping $\mathbb{N}_\infty$ with an order. For a very simple example, order $\mathbb{N}_\infty$ as the chain $\mathbb{N}$ with ∞ adjoined as top element. Take $x \not\geqslant y$. Then $y > x$ and $\downarrow x$, which is clopen because it is finite and does not contain ∞, contains x but not y. Hence we have a CTOD space; its lattice of clopen down-sets is isomorphic to the chain $\mathbb{N} \oplus \mathbf{1}$.

Alternatively, consider the ordered space Y obtained by equipping $\mathbb{N}_\infty$ with the order depicted in Figure 10.2. We have $n \succ n - 1$ and $n \succ n + 1$ for each even n.

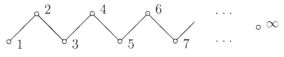

Figure 10.2

For each $n \in \mathbb{N}$, the down-set $\downarrow n$ is finite and does not contain ∞ and so is clopen. Given $x \not\geqslant y$ in Y, we claim that there exists

$U \in \mathcal{O}(Y)$ such that $x \in U$ and $y \notin U$. Either $x \neq \infty$, in which case $y \notin {\downarrow} x$ and we may take $U = {\downarrow} x$, or $x = \infty$, in which case we may take $U = \{u \in Y \mid u \notin \{1, 2, ..., 2y\}\}$. Hence Y is a CTOD space. The associated lattice $\mathcal{O}(Y)$—a sublattice of $FC(\mathbb{N})$—is easily described.

These examples illustrate how CTOD spaces can be constructed by imposing suitable order relations on a given Boolean space. Alternatively, we might start from an ordered set and try to make it into a CTOD space by topologizing it. This raises the interesting, but difficult, problem of **representability**: when is a given ordered set isomorphic to $\langle \mathcal{I}_p(L); \subseteq \rangle$ for some bounded distributive lattice L, or, more generally, an arbitrary distributive lattice L? For more information on this, see [51].

(3) Let C be the Cantor 'middle third' set, regarded as a subset of $[0, 1]$. Then C is compact, since C is obtained from $[0, 1]$ by removing open intervals. Also, if $x < y$ in C, there exists u such that $x < u < y$ and $u \notin C$. Then $C \cap [0, u]$ is clopen, contains x and does not contain y. It follows that C is a Boolean space (in fact with the order inherited from $\mathbb{R}$ it is a CTOD space). Those reasonably adept at topology can now prove that $\mathcal{P}(C)$ is an example of a countable Boolean algebra with no atoms.

(4) Let S be a set. As a subbasis for a topology on $\langle \wp(S); \subseteq \rangle$ we take the collection of up-sets of the form ${\uparrow}\{s\}$ (for $s \in S$), and their complements. Alexander's Subbasis Lemma can be used to prove, much as in the proof of 10.14, that $\wp(S)$ then becomes a CTOD space. When S is countable, the underlying Boolean space is homeomorphic to the Cantor space described above. For further information on this example see [45].

10.22 Distributive lattices and CTOD spaces in partnership. Denote the class of bounded distributive lattices by **D**, and the class of compact totally order-disconnected spaces by **P**. Define maps D and E by

$$D: L \mapsto \mathcal{I}_p(L) \qquad (L \in \mathbf{D}),$$
$$E: X \mapsto \mathcal{O}(X) \qquad (X \in \mathbf{P}).$$

Theorems 10.18 and 10.19, assert that, for all $L \in \mathbf{D}$ and $X \in \mathbf{P}$,

$$ED(L) \cong L \quad \text{and} \quad DE(X) \cong X;$$

the latter $\cong$ means 'is order-homeomorphic to'.

We may use the isomorphism between L and $ED(L)$ to represent the members of **D** concretely as lattices of the form $\mathcal{O}(X)$ for $X \in \mathbf{P}$.

As an immediate application we note that the representation allows us to construct a 'smallest' Boolean algebra B containing (an isomorphic copy of) a given lattice $L \in \mathbf{D}$: identify L with $\mathcal{O}(X)$ and take $B = \mathcal{P}(X)$. Lemma 10.17 shows how $\mathcal{O}(X)$ and $\mathcal{P}(X)$ are related.

As in the finite case, X is generally very much simpler than $\mathcal{O}(X)$. We can relate properties of $\mathcal{O}(X)$ to properties of X, as in 8.21 and 8.22. To complement Chapter 8, we relegate the generalization of these results to the exercises, and use as illustrations in the text a discussion of pseudocomplements (the subject of an exercise in the finite case) and of ideals (of no interest in the finite case since every ideal is then principal).

Before we proceed, a comment on a conflict of notation is called for. We have customarily used lower case letters $a, b, c, \ldots$ for lattice elements. On the other hand, when $L \in \mathbf{D}$ is concretely represented as the lattice $\mathcal{O}(X)$ of clopen down-sets of its dual space X it is natural to denote subsets of X, including elements of L, by $U, V, W, \ldots$, and the points of the space X by $x, y, z, \ldots$. The problem is made worse by the fact that the points of a prime ideal space are ideals and, as sets, ideals are usually denoted by upper case letters. We perforce display a kind of notational schizophrenia and switch between these conflicting notational styles as the context indicates.

10.23 Pseudocomplements. We have encountered many distributive lattices which are not Boolean: the requirement that every element a of a bounded lattice L have a complement is stringent. There are many ways to weaken the condition. One possibility is to define the **pseudocomplement** of an element a in a lattice L with 0 to be

$$a^* = \max\{\, b \in L \mid b \wedge a = 0 \,\},$$

if this exists. Pseudocomplements in finite lattices were the subject of Exercise 8.22. Now consider $L = \mathcal{O}(X)$, where X is a CTOD space. When does $U \in L$ have a pseudocomplement? We claim that this is so if and only if $\uparrow U$ is clopen, and that then $U^* = X \smallsetminus \uparrow U$. To prove the claim, first observe that a down-set W in X does not intersect U if and only if $W \subseteq X \smallsetminus \uparrow U$. Hence $X \smallsetminus \uparrow U$ is the largest down-set disjoint from U and, if it is also clopen, it must be U^*. Conversely, assume U^* exists. Take $x \notin \uparrow U$. We show $x \in U^*$, from which it follows that $U^* = X \smallsetminus \uparrow U$. Exercise 10.9 implies $\uparrow U$ is closed. By the dual of Lemma 10.16(ii) we can find a clopen up-set V such that $x \notin V$ and $\uparrow U \subseteq V$. Then $(X \smallsetminus V) \cap U = \varnothing$, so $X \smallsetminus V \subseteq U^*$ by definition of U^*. This implies $x \in U^*$. See Figure 10.3.

10.24 Duality for ideals. Let $L = \mathcal{O}(X)$ where X is a CTOD space whose family of open down-sets we denote by $\mathcal{L}$. How can we describe

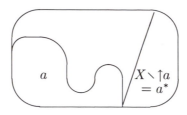

Figure 10.3

the ideals and filters of L in terms of X? An ideal J is determined by its members, which are clopen down-sets of X. Define

$$\Phi(J) = \bigcup \{U \mid U \in J\} \quad \text{(for } J \in \mathcal{I}(L)\text{)};$$

as a union of clopen sets, $\Phi(J)$ is an open set (but not in general clopen). In the other direction, define

$$\Psi(W) = \{U \in \mathcal{O}(X) \mid U \subseteq W\} \quad \text{(for } W \in \mathcal{L}\text{)};$$

it is easily checked that $\Psi(W)$ is an ideal of L. Further, we claim that

$$\Phi(\Psi(W)) = W \text{ for all } W \in \mathcal{L} \quad \text{and} \quad \Psi(\Phi(J)) = J \text{ for all } J \in \mathcal{I}(L).$$

The first equation asserts that an open down-set W is the union of the clopen down-sets contained in it. To prove this, take any $x \in W$ and use the dual of Lemma 10.16(ii) to find a clopen up-set V containing the closed up-set $X \smallsetminus W$ but with $x \notin V$; then $x \in X \smallsetminus V$, a clopen down-set inside W (see Figure 10.4).

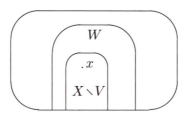

Figure 10.4

Certainly $J \subseteq \Psi(\Phi(J))$ for each ideal J. Take $V \in \Psi(\Phi(J))$. This means that J, regarded as a family of open subsets of X, is an open cover of the clopen set V. By 10.7A, only finitely many elements of J are needed to cover V, say $U_{x_1}, \ldots, U_{x_n}$. But $V \subseteq U_{x_1} \cup \cdots \cup U_{x_n}$ implies $V \in J$, since J is an ideal. This establishes the second equation.

The bijective correspondence we have set up between $\mathcal{I}(L)$ and $\mathcal{L}$ is in fact a lattice isomorphism (an easy exercise). In addition, special types of ideal correspond to special types of open set (see Exercise 10.13).

Filters may be treated similarly: $\mathcal{F}(L) \cong \mathcal{F}$, the lattice of open up-sets of X.

10.25 Duality. We have amassed a lot of evidence that lattice concepts in **D** can be translated into ordered set concepts in **P** and vice versa. This 'D-P dictionary' results from there being what is known as a **(full) duality** between **D** (bounded distributive lattices + $\{0, 1\}$-homomorphisms) and **P** (CTOD spaces + continuous order-preserving maps). In line with the philosophy in 1.16, we have here extended the scope of the symbols **D** and **P** to encompass structure-preserving maps as well as objects. For $L, K \in \mathbf{D}$ and $X, Y \in \mathbf{P}$, we denote the set of $\{0, 1\}$-homomorphisms from L to K by $\mathbf{D}(L, K)$ and the set of continuous order-preserving maps from Y to X by $\mathbf{P}(Y, X)$.

The way the duality is required to work is formally laid out in (O) and (M) below. Note the reversal of the directions in (M). We do not digress to discuss the categorical background to these conditions, but do point out that those familiar with dual vector spaces should have a sense of *déjà vu*.

(O) There exist maps $D: \mathbf{D} \to \mathbf{P}$ and $E: \mathbf{P} \to \mathbf{D}$ such that

    (i) for each $L \in \mathbf{D}$, there exists $\eta_L: L \to ED(L)$ such that η_L is an isomorphism,

    (ii) for each $X \in \mathbf{P}$, there exists $\varepsilon_X: X \to DE(X)$ such that ε_X is an order-homeomorphism.

(M) For any $L, K \in \mathbf{D}$, there exists for each $f \in \mathbf{D}(L, K)$ a map $D(f) \in \mathbf{P}(D(K), D(L))$. For each $X, Y \in \mathbf{P}$, there exists for $\varphi \in \mathbf{P}(Y, X)$ a map $E(\varphi) \in \mathbf{D}(E(X), E(Y))$. The maps $D: \mathbf{D}(L, K) \to \mathbf{P}(D(K), D(L))$ and $E: \mathbf{P}(Y, X) \to \mathbf{D}(E(X), E(Y))$ are bijections and the diagrams below commute.

$$
\begin{array}{ccc}
L & \xrightarrow{\ f\ } & K \\
\eta_L \downarrow & & \downarrow \eta_K \\
ED(L) & \xrightarrow{ED(f)} & ED(K)
\end{array}
\qquad
\begin{array}{ccc}
Y & \xrightarrow{\ \varphi\ } & X \\
\varepsilon_Y \downarrow & & \downarrow \varepsilon_X \\
DE(Y) & \xrightarrow{DE(\varphi)} & DE(X)
\end{array}
$$

By Exercise 10.2, $\mathbf{P}(X, \mathbf{2})$ is a $\{0, 1\}$-sublattice of $\mathbf{2}^X$ and is isomorphic to $E(X)^\partial$. By Exercise 9.2, $\mathbf{D}(L, \mathbf{2})$ is order-isomorphic to $D(L)^\partial$. Exercise 10.18 indicates that these 'homsets' provide a viable alternative for setting up the duality between $\mathbf{D}$ and $\mathbf{P}$. In fact, the homset approach is technically superior but is inappropriate for a first pass at this topic as it lacks the geometric appeal inherent in the approach via prime ideals and clopen down-sets.

We already have (O), the object part of the duality, with η_L and ε_X the maps supplied by Theorems 10.18 and 10.19 appropriately labelled. Theorem 10.26 provides (M), the correspondence for maps.

10.26 Theorem. Let $L, K \in \mathbf{D}$ and $X, Y \in \mathbf{P}$. Then to each map $f \in \mathbf{D}(L, K)$ is associated a map $D(f) \in \mathbf{P}(D(K), D(L))$ and to each map $\varphi \in \mathbf{P}(Y, X)$ is associated a map $E(\varphi) \in \mathbf{D}(E(X), E(Y))$. The maps $D: \mathbf{D}(L, K) \to \mathbf{P}(D(K), D(L))$ and $E: \mathbf{P}(D(K), D(L)) \to \mathbf{D}(L, K)$ are bijections and

$$a \in (D(f))(y) \Longleftrightarrow f(a) \in y \quad \text{for all } a \in L, \, y \in Y.$$

Further,

(i) f *is one-to-one if and only if* $D(f)$ *is onto;*

(ii) f *is onto if and only if* $D(f)$ *is an order-embedding.*

Proof. It is elementary to show that, given $\varphi \in \mathbf{P}(Y, X)$, the formula $(E(\varphi))(U) := \varphi^{-1}(U)$ for $U \in E(X)$ defines a $\{0, 1\}$-homomorphism $E(\varphi): E(X) \to E(Y)$.

Now assume $f \in \mathbf{D}(L, K)$. Take $y \in Y$ (so y is a prime ideal of K) and define $(D(f))(y) := f^{-1}(y)$. It is routine to check that $(D(f))(y)$ is a prime ideal in L and that $D(f)$ is order-preserving. To prove $D(f)$ is continuous it is enough (see the proof of 10.19) to prove that $D(f)^{-1}(\eta_L(a))$ is clopen for each $a \in L$. (Here $\eta_L(a)$ is just the set we previously wrote as X_a; we have used the alternative notation because we now have to keep track of subbasic clopen sets in two different dual spaces.) We have

$$y \in D(f)^{-1}(\eta_L(a)) \Longleftrightarrow (D(f))(y) \in X_a \Longleftrightarrow a \notin (D(f))(y)$$
$$\Longleftrightarrow f(a) \notin y \Longleftrightarrow y \in \eta_K(f(a)),$$

and $\eta_K(f(a))$ is clopen in $D(K)$. The maps D and E are therefore well defined. To verify that diagrams in (M) indeed commute requires a fairly energetic definition-chase, which we leave to the reader, along with the proofs of (i) and (ii) (Exercises 10.6 and 10.7). ∎

Congruences play a very central role in the more advanced theory of lattices. In particular they are important in the study of distributive lattices with additional operations. It is therefore gratifying that the lattice of congruences of $L \in \mathbf{D}$ turns out to have a very nice concrete representation under our duality.

10.27 Duals of congruences. Let $L \in \mathbf{D}$. As usual, our work is simplified by assuming that $L = \mathcal{O}(X)$ for some $X \in \mathbf{P}$. We have correspondences

congruences on $L \leftrightarrow \{0, 1\}$-homomorphisms with domain L (by 5.21)

$$\leftrightarrow \text{ continuous order-embeddings into } X \text{ (by 10.26)}$$

$$\leftrightarrow \text{ closed subsets of } X \text{ (by 10.7A).}$$

Further, a larger congruence on L gives a smaller homomorphic image and hence a smaller closed subset of X. We would therefore expect the lattice $\operatorname{Con} L$ to be isomorphic to the dual of the lattice $\Gamma(X)$ of closed subsets of X. We now elucidate how this isomorphism works, leaving the reader to check the details (not difficult, but a clear head is recommended).

The first step is easy. For each closed subset Y of X, we can define a congruence $\theta(Y)$ on $\mathcal{O}(X)$ by

$$(U, V) \in \theta(Y) \iff U \cap Y = V \cap Y$$

for $U, V \in \mathcal{O}(X)$. The congruence $\theta(Y)$ is the kernel of the $\{0, 1\}$-homomorphism f from $\mathcal{O}(X)$ onto $\mathcal{O}(Y)$ given by $f(U) := U \cap Y$.

Now let θ be any congruence on L and let $q \colon \mathcal{O}(X) \to \mathcal{O}(X)/\theta$ be the natural quotient map. Theorems 5.20 and 10.18 guarantee that there exists an isomorphism $h \colon \mathcal{O}(X)/\theta \to \mathcal{O}(Z)$ for some $Z \in \mathbf{P}$. Then $h \circ q$ maps $\mathcal{O}(X)$ onto $\mathcal{O}(Z)$. By 10.26 the dual of this map is an order-embedding $\varphi \colon Z \hookrightarrow X$. The set $Y := \varphi(Z)$ is closed in X and $\theta = \theta(Y)$.

This shows that $Y \mapsto \theta(Y)$ is a map from $\Gamma(X)$ onto $\operatorname{Con} L$. It is an order-isomorphism between $\Gamma(X)^\partial$ and $\operatorname{Con} L$ provided $\theta(Y_1) \subseteq \theta(Y_2)$ implies $Y_1 \supseteq Y_2$ in $\Gamma(X)$ (the reverse implication is trivial). It is enough to prove that whenever Z is a closed subset of X and $y \in X \smallsetminus Z$, there exist clopen down-sets U, V in X such that $U \cap Z = V \cap Z$ and $y \in V \smallsetminus U$. To do this, apply Lemma 10.16(ii) twice: first to $\downarrow(Z \cap \downarrow y)$ and $\uparrow y$ to yield U, then to $U \cup \downarrow y$ and $\uparrow(Z \smallsetminus U)$ to yield V.

Hence $\Gamma(X)^\partial \cong \operatorname{Con} \mathcal{O}(X)$, via the map $Y \mapsto \theta(Y)$. We deduce that $\operatorname{Con} \mathcal{O}(X)$ is isomorphic to the lattice of open subsets of X.

Appendix: A topological toolkit

This appendix provides a very concise summary of the results from topology needed in Chapter 10 and its exercises. Our account aims solely to pinpoint those topological ideas we need. Any standard text may be consulted for proofs and further motivation. Our references are to W.A. Sutherland, *Introduction to metric and topological spaces* [17].

Topology is usually introduced as an abstraction of concepts first met in elementary analysis, such as open neighbourhood and continuous function. In a topological space, a family of **open sets** generalizes the open neighbourhoods of the euclidean spaces $\mathbb{R}^n$. The axioms for a topological space bring under topology's umbrella many structures which are very unlike euclidean spaces. It is certain such spaces that concern us. The metric spaces which utilize the idea of a distance function analogous to the modulus function on $\mathbb{R}$ and which are frequently used as a stepping stone to topological spaces play no role here.

Proving results in topology demands a certain facility in manipulating sets and maps. The formulae set out in [17], pp. xi-xiii, are a necessary stock in trade.

10.1A Topological spaces.

A **topological space** $(X; \mathcal{T})$ consists of a set X and a family $\mathcal{T}$ of subsets of X such that

(T1) $\emptyset \in \mathcal{T}$ and $X \in \mathcal{T}$;

(T2) a finite intersection of members of $\mathcal{T}$ is in $\mathcal{T}$;

(T3) an arbitrary union of members of $\mathcal{T}$ is in $\mathcal{T}$.

The family $\mathcal{T}$ is called a **topology** on X and the members of $\mathcal{T}$ are called **open sets**. We write X in place of $(X; \mathcal{T})$ where $\mathcal{T}$ is the only topology under consideration. The justification for the lopsidedness of conditions (T2) and (T3) is 'because it works'. The standard topology $\mathcal{T}_\mathbb{R}$ on $\mathbb{R}$ consists of

$$\{\, U \subseteq \mathbb{R} \mid (\forall x \in U)(\exists \delta > 0)\ (x - \delta, x + \delta) \subseteq U \,\};$$

here δ may depend on x. Equivalently, $\mathcal{T}_\mathbb{R}$ consists of those sets which can be expressed as unions of open intervals, together with $\emptyset$. Certainly this family does satisfy (T1), (T2) and (T3). The equation $\bigcap_{n \geqslant 1}(-1/n, 1/n) = \{0\}$ exhibits an intersection of open sets which is not open. Thus (T2) could not be strengthened to require arbitrary intersections of open sets to be open without sacrificing the motivating example of the 'usual' open sets of $\mathbb{R}$.

Given a topological space $(X; \mathcal{T})$ we define a subset of X to be **closed** if it belongs to $\Gamma(X) := \{\, X \setminus U \mid U \in \mathcal{T} \,\}$. The family $\Gamma(X)$ is closed under arbitrary intersections and finite unions. For every $A \subseteq X$ there exists a smallest closed set $\overline{A}$ containing A; see 2.20 and [17], 3.7.

Sets which are both open and closed are called **clopen**. Very many of the topological spaces encountered in elementary analysis and geometry are **connected**, in the sense that their only clopen subsets are the whole space and the empty set. The spaces used in our representation theory, by contrast, have an ample supply of clopen sets. For a discussion of connectedness, see [17], Chapter 6.

10.2A Subspaces [[17], 3.4]. Let $(X; \mathcal{T})$ be a topological space. Any subset Y of X inherits a topology in a natural way. It is given by

$$\mathcal{T}_Y := \{ V \subseteq Y \mid V = U \cap Y \text{ for some } U \in \mathcal{T} \}.$$

10.3A Bases and subbases [[17], 3.2, 3.3]. We need to be able to create a topology on a set X in which a specified family $\mathcal{S}$ of subsets of X are open sets. We shall always asssume that $\mathcal{S}$ contains $\varnothing$ and X. If $\mathcal{S}$ is already closed under finite intersections, then we define $\mathcal{T}$ to be those sets which are unions of sets in $\mathcal{S}$. Then $\mathcal{T}$ satisfies (T1), (T2) and (T3) and $\mathcal{S}$ is said to be a **basis** for $\mathcal{T}$. In general, to obtain a topology containing $\mathcal{S}$ we first form $\mathcal{B}$, the family of sets which are finite intersections of members of $\mathcal{S}$, and then define $\mathcal{T}$ to be all arbitrary unions of members of $\mathcal{B}$. In this case $\mathcal{S}$ is called a **subbasis** for $\mathcal{T}$.

10.4A Continuity. Let $(X; \mathcal{T})$ and $(X'; \mathcal{T}')$ be topological spaces and $f : X \to X'$ a map. Then [[17], 3.7.8, 3.2.5 and p. xii] the following conditions are equivalent:

(i) $f^{-1}(U)$ is open in X whenever U is open in X';

(i)' $f^{-1}(V)$ is closed in X whenever V is closed in X';

(ii) $f^{-1}(U)$ is open in X for every $U \in \mathcal{S}$, where $\mathcal{S}$ is a given basis or subbasis for $\mathcal{T}'$.

When f satisfies any of these conditions it is said to be **continuous**. In the special case that $(X; \mathcal{T}) = (X'; \mathcal{T}') = (\mathbb{R}; \mathcal{T}_\mathbb{R})$ and $\mathcal{S}$ is the family of subintervals (a, b) (for $-\infty < a < b < \infty$), plus $\mathbb{R}$ and $\varnothing$, (ii) is just a restatement of the ε-δ definition of continuity that so plagues students.

The map $f : X \to X'$ is said to be a **homeomorphism** if f is bijective and both f and f^{-1} are continuous. Homeomorphisms are topology's isomorphisms.

10.5A Hausdorff spaces [[17], Chapter 4]. Of a hierarchy of possible separation conditions, augmenting the topological space axioms, the most important to us is the Hausdorff condition. The topological space

$(X; T)$ is said to be **Hausdorff** if, given $x, y \in X$ with $x \neq y$, there exist open sets U_1, U_2 such that $x \in U_1, y \in U_2$ and $U_1 \cap U_2 = \varnothing$. Mnemonically, X is Hausdorff if distinct points can be 'housed off' in disjoint open sets. It is easy to prove the useful result that singleton sets in a Hausdorff space are closed.

10.6A Compactness [[17], Chapter 5]. A prime objective of elementary topology is to set in their wider topological context the various results concerning closed bounded subintervals of $\mathbb{R}$ and the continuous real-valued functions on them. The famous Heine–Borel Theorem states that a subset of $\mathbb{R}$ is closed and bounded if and only if it is compact, in the sense we shortly define. Compactness is a fundamental topological concept and may be regarded as a substitute for finiteness. It frequently compensates for the restriction to finite intersections in axiom (T2) by allowing arbitrary families of open sets to be reduced to finite families. All the spaces we use in our representation theory are compact.

Let $(X; T)$ be a topological space and let $\mathcal{U} := \{U_i\}_{i \in I} \subseteq T$. The family $\mathcal{U}$ is called an **open cover** of $Y \subseteq X$ if $Y \subseteq \bigcup_{i \in I} U_i$. A finite subset of $\mathcal{U}$ whose union still contains Y is a **finite subcover**. We say Y is **compact** if every open cover of Y has a finite subcover.

The lemmas below contain basic results about compact spaces which are also Hausdorff. The first relates compactness and closedness and shows that continuous maps behave well.

10.7A Lemma. *Let $(X; T)$ be a compact Hausdorff space.*

(i) [[17], 5.4.2, 5.6.1] *A subset Y of X is compact if and only if it is closed.*

(ii) [[17], 5.5.1, 5.9.1] *Let $f: X \to X'$ be a continuous map, where $(X'; T')$ is any topological space.*

  (a) *$f(X)$ is a compact subset of X'.*

  (b) *If $(X'; T')$ is Hausdorff and $f: X \to X'$ is bijective, then f is a homeomorphism.*

Since the next lemma is given in [17] only as an exercise we outline its proof. The lemma strengthens the Hausdorff condition, which is recaptured by taking the closed sets to be singletons.

10.8A Lemma. *Let $\langle X; T \rangle$ be a compact Hausdorff space.*

(i) *Let V be a closed subset of X and $x \notin V$. Then there exist disjoint open sets W_1 and W_2 such that $x \in W_1$ and $V \subseteq W_2$.*

(ii) *Let V_1 and V_2 be disjoint closed subsets of X. Then there exist disjoint open sets U_1 and U_2 such that $V_i \subseteq U_i$ for $i = 1, 2$.*

Proof. (i) For $y \in V$, use the Hausdorff condition to construct disjoint open sets $U_1^{x,y}$ and $U_2^{x,y}$ containing x and y respectively. Then $\mathcal{U}_2 := \{U_2^{x,y} \mid y \in V\}$ is an open cover of V, which is compact by 10.7A. Take a finite subcover $\{U_2^{x,y_j} \mid j = 1, \ldots, n\}$. Let $U_1^x := \bigcap_{1 \leqslant j \leqslant n} U_1^{x,y_j}$ and $U_2^x := \bigcup_{1 \leqslant j \leqslant n} U_2^{x,y_j}$. Then U_1^x and U_2^x are disjoint, since each U_2^{x,y_j} does not intersect the corrsponding U_1^{x,y_j} and so is disjoint from U_1^x. Also U_1^x and U_2^x are open. (Note how compactness reduces the intersection we need to a finite one, so (T2) applies.) These sets contain x and V respectively. Take $W_1 := U_1^x$ and $W_2 := U_2^x$ to obtain (i).

For (ii) we repeat the process, taking $V := V_2$ and letting x vary over V_1. The family $\mathcal{U}_1 := \{U_1^x \mid x \in V_1\}$ is an open cover of the compact set V_1. Take a finite subcover $\{U_1^{x_i} \mid i = 1, \ldots, m\}$ and define $U_1 := \bigcup_{1 \leqslant i \leqslant m} U_1^{x_i}$ and $U_2 := \bigcap_{1 \leqslant i \leqslant m} U_2^{x_i}$. ∎

The last of this group of lemmas enables us to fit our finite representation theory into the general theory in Chapter 10.

10.9A Lemma. *Let $\langle X; \mathcal{T} \rangle$ be a compact Hausdorff space. Then the following conditions are equivalent:*

(i) *X is finite;*

(ii) *every subset of X is open (that is, $\mathcal{T}$ is discrete);*

(iii) *every subset of X is clopen.*

Proof. Trivially (ii) $\Leftrightarrow$ (iii). To prove (iii) $\Rightarrow$ (i) consider the open cover $\{\{x\} \mid x \in X\}$. Finally assume (i). For $\varnothing \neq Y \subseteq X$, the set $X \smallsetminus Y$ is a finite union of singleton sets, which are closed because X is Hausdorff. So $X \smallsetminus Y$ is closed, whence Y is open. ∎

The deepest result about compact spaces we need is Alexander's Subbasis Lemma. We prove this using (BPI); recall 9.12. We have elected to avoid the machinery of product spaces in the main text. Those who already know about product spaces will realize that Alexander's Lemma is closely related to Tychonoff's Theorem. The latter is employed directly in the alternative approach to duality outlined in Exercise 10.18.

10.10A Alexander's Subbasis Lemma. *Let $(X; \mathcal{T})$ be a topological space and $\mathcal{S}$ a subbasis for $\mathcal{T}$. Then X is compact if every open cover of X by members of $\mathcal{S}$ has a finite subcover.*

Proof. Let $\mathcal{B}$ be the basis formed from all finite intersections of members of $\mathcal{S}$. To prove X is compact it is enough to show that a given open cover $\mathcal{U}$ of X by sets in $\mathcal{B}$ has a finite subcover. Suppose this is false. Define J to be the ideal in $\wp(X)$ generated by $\mathcal{U}$, so a typical element of J is a subset of $U_1 \cup \cdots \cup U_k$ for some $U_1, \ldots, U_k \in \mathcal{U}$ (see 9.1); J

is proper, by our hypothesis. Use (BPI) to construct a prime ideal I in $\wp(X)$ containing J. For each $x \in X$, there exists $U(x) \in \mathcal{U}$ with $x \in U(x)$. Each $U(x)$ is a finite intersection of members of $\mathcal{S}$ and belongs to I since $\mathcal{U} \subseteq I$. As I is prime we may assume that $U(x)$ itself lies in $\mathcal{S}$. Let $\mathcal{V} := \{ U(x) \mid x \in X \}$. Then $\mathcal{V}$ is an open cover of X by members of $\mathcal{S}$ and so by assumption has a finite subcover. But then $X = U(x_1) \cup \cdots \cup U(x_n)$ for some finite subset $\{x_1, \ldots, x_n\}$ of X, so that $X \in I$, ⨍. ∎

Exercises

Exercises from the text. Prove (or complete the proof of) each of the following Lemmas: 10.1, 10.10(ii), 10.16, 10.17. Show that the family $\mathcal{T}$ of subsets of N_0, defined in Example 10.11, is a topology on N_0—but beware special cases when checking the conditions (T1)–(T3)!

Prove the assertion in the penultimate paragraph of 10.24 that $\mathcal{I}(L) \cong \mathcal{L}$. Fill in the details in 10.27.

10.1 A topological space X is called **zero-dimensional** if the clopen subsets of X form a basis for the topology. Prove that the following conditions are equivalent:

    (i) X is a Boolean space;

    (ii) X is compact, Hausdorff and the only connected subsets of X are the singletons $\{x\}$ for $x \in X$.

    (iii) X is compact, T_0 (see Exercise 1.25) and zero-dimensional.

10.2 Let X be a CTOD space.

    (i) Assume that $\varphi \in \mathbf{P}(X, \mathbf{2})$. Show that $\varphi^{-1}(0)$ is a clopen down-set in X.

    (ii) Let U be a clopen down-set in X. Define $\varphi \colon X \to \mathbf{2}$ by

$$\varphi(x) = \begin{cases} 1 & \text{if } x \notin U, \\ 0 & \text{if } x \in U. \end{cases}$$

    Show that $\varphi \in \mathbf{P}(X, \mathbf{2})$.

    (iii) Let L denote the set $\mathbf{P}(X, \mathbf{2})$. Show that L is a $\{0, 1\}$-sublattice of $\mathbf{2}^X$. Set up an order-isomorphism between $\mathcal{O}(X)$ and L^{∂}.

10.3 Let X and Y be topological spaces with $X \cap Y = \varnothing$. There is a natural topology on $X \cup Y$ whose open sets are of the form $A \dot\cup B$ where A is open in X and B is open in Y. Thus $X \dot\cup Y$ is an ordered space if both X and Y are. This ordered space is called the **disjoint union** of X and Y. (If X and Y are not disjoint, we must first replace them by disjoint copies—see Exercise 1.21 where the corresponding construction was considered for ordered sets.)

(i) Show that if X and Y are (disjoint) Boolean spaces then $X \mathbin{\dot\cup} Y$ is also a Boolean space.

(ii) Show that if X and Y are (disjoint) CTOD spaces then $X \mathbin{\dot\cup} Y$ is also a CTOD space.

(iii) (a) Show that, if $X, Y \in \mathbf{P}$, then $\mathcal{O}(X \mathbin{\dot\cup} Y) \cong \mathcal{O}(X) \times \mathcal{O}(Y)$.

    (b) Let $L, K \in \mathbf{D}$. Use (a) and the duality between $\mathbf{D}$ and $\mathbf{P}$ to show there is an order-homeomorphism from the space $D(L \times K)$ of prime ideals of $L \times K$ to $D(L) \mathbin{\dot\cup} D(K)$. (Compare this with Exercise 9.3.)

10.4 Given $X, Y \in \mathbf{P}$, we denote by $X \oplus Y$ the usual linear sum of the underlying ordered sets endowed with the disjoint union topology defined in Exercise 10.3.

(i) Show that $X \oplus Y \in \mathbf{P}$.

(ii) Show that $\mathcal{O}(X \oplus Y) \cong \mathcal{O}(X) \overline{\oplus} \mathcal{O}(Y)$, where $\overline{\oplus}$ denotes the vertical sum as defined in Exercise 8.9 (the finite case).

(iii) Let $L, K \in \mathbf{D}$. Find all prime ideals in $L \oplus K$. Hence prove, without use of the duality between $\mathbf{D}$ and $\mathbf{P}$, that $D(L \oplus K)$ is order-homeomorphic to $D(L) \oplus \mathbf{1} \oplus D(K)$.

10.5 Consider the ordered spaces shown in Figure 10.5. In each case the order and the topology should be apparent. For example, in X_1 the only comparabilities are $a_1 \prec b_1$ and $a_n \prec b_{n-1}$, $a_n \prec b_n$ for $n \geqslant 2$; note, in particular, that $a_\infty \parallel b_\infty$. As a topological space, X_1 is the disjoint union of two copies on $\mathbb{N}_\infty$, namely $\{a_1, a_2, \dots\} \cup \{a_\infty\}$ and $\{b_1, b_2, \dots\} \cup \{b_\infty\}$. The other examples are built similarly from one or two copies of $\mathbb{N}_\infty$.

(i) Show that $X_1 \notin \mathbf{P}$.

(ii) Show that $X_2 \in \mathbf{P}$ and describe all the elements of $\mathcal{O}(X_2)$.

(iii) Consider the CTOD space Y given in Figure 10.2. Show that $\mathcal{O}(Y)$ is a sublattice of $\mathrm{FC}(\mathbb{N})$ and describe the elements of $\mathcal{O}(Y)$ (in terms of odd and even numbers).

(iv) Show that $X_3 \in \mathbf{P}$. Show that $\mathcal{O}(X_3)$ is isomorphic to a sublattice of $\mathrm{FC}(\mathbb{N}) \times \mathrm{FC}(\mathbb{N})$. [Hint. Find a continuous order-preserving map from $\mathbb{N}_\infty \mathbin{\dot\cup} \mathbb{N}_\infty$ onto X_3 then use the duality.] Give an explicit description of the elements of this sublattice of $\mathrm{FC}(\mathbb{N}) \times \mathrm{FC}(\mathbb{N})$.

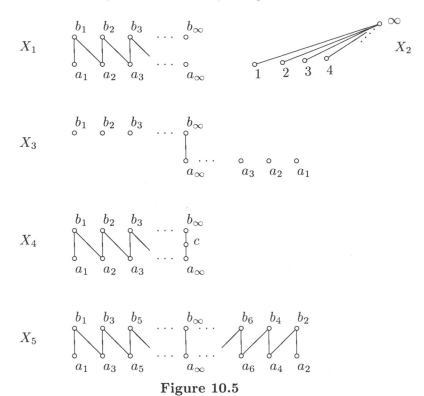

Figure 10.5

(v) Show that $X_4 \in \mathbf{P}$ and $X_5 \in \mathbf{P}$.

10.6 This exercise completes the proof of Theorem 10.26. Let $L, K \in \mathbf{D}$
and let $f \in \mathbf{D}(L, K)$.

(i) (a) Show that, if I is a prime ideal of K, then $f^{-1}(I)$ is a
prime ideal of L.

(b) Show that, if K_1 is a sublattice of K and J is an ideal of
K_1, then $\downarrow J$ is an ideal of K.

(ii) (a) Let K_1 be a sublattice of K and let I_1 be a prime ideal of
K_1. Show that there exists a prime ideal I of K satisfying
$K_1 \cap I = I_1$ [Hint. Apply (DPI) to the ideal $\downarrow I$ and filter
$\uparrow(K_1 \setminus I_1)$ of K.]

(b) Prove that, if f is one-to-one, then $D(f)$ is onto.

(c) Prove that, if $D(f)$ is onto, then f is one-to-one. [Hint.
Prove the contrapositive. Assume that $a, b \in L$ satisfy
$a \neq b$ and $f(a) = f(b)$. Use Theorem 10.3 to obtain a
prime ideal $I \in D(L)$ which is not in the image of $D(f)$.]

(iii) (a) Show that, if f is onto, then $D(f)$ is an order-embedding.

(b) Prove that, if $D(f)$ is an order-embedding, then f is onto. [Hint. Again prove the contrapositive. Let $K_1 = f(L)$ and assume that $a \in K \setminus K_1$. Apply (DPI) first to the pair $\downarrow a$, $\uparrow(K_1 \cap \uparrow a)$ to yield a prime ideal I, and then to the pair $\downarrow(K_1 \cap I)$, $\uparrow a$ to yield a second prime ideal J. Show that $D(f)(I) \subseteq D(f)(J)$ while $I \not\subseteq J$.]

10.7 Let $X, Y \in \mathbf{P}$, let X_1 be a closed subset of X and let $\varphi \in \mathbf{P}(Y, X)$.

(i) Let U_1 be a clopen down-set in X_1. Prove that there exists a clopen down-set U in X satisfying $X_1 \cap U = U_1$. [Hint. Apply Lemma 10.16(ii) to the pair $\downarrow U_1$, $\uparrow(X_1 \setminus U_1)$.]

(ii) Prove that, if φ is an order-embedding, then $E(\varphi)$ is onto.

(iii) Use (ii) and the duality between $\mathbf{D}$ and $\mathbf{P}$ to give an alternative proof of Exercise 10.6(iii). [Hint. Use the left-hand commutative diagram in 10.25.]

10.8 A lattice $M \in \mathbf{D}$ is said to be **injective in** $\mathbf{D}$ if, for every $K \in \mathbf{D}$ and every $\{0, 1\}$-embedding $f \colon L \to K$, each $\{0, 1\}$-homomorphism $g \colon L \to M$ can be extended to a homomorphism $\bar{g} \colon K \to M$, that is $\bar{g}(f(a)) = g(a)$ for all $a \in L$. Similarly, a CTOD space Z is **injective in** $\mathbf{P}$ if, for every $X \in \mathbf{P}$ and every continuous order-embedding $\varphi \colon Y \to X$, each continuous order-preserving map $\psi \colon Y \to Z$ extends to a continuous order-preserving map $\bar{\varphi} \colon X \to Z$.

(i) Use Exercises 10.6(ii)(a) and 9.2 to show that $\mathbf{2}$ is injective in $\mathbf{D}$.

(ii) Use Exercises 10.2 and 10.7(i) to show that $\mathbf{2}$ is injective in $\mathbf{P}$.

10.9 Let X be a CTOD space.

(i) Show that $\downarrow y$ and $\uparrow y$ are closed for each $y \in X$.

(ii) Show that, if $Y \subseteq X$ is closed in X, then $\uparrow Y$ and $\downarrow Y$ are closed in X.

(iii) Show that Y is a closed down-set in X if and only if Y is an intersection of clopen down-sets.

(iv) Prove that every directed subset D of X has a join in X. [Hint. Use the duality.] (Hence X is a pre-CPO.)

10.10 For any ordered set P, let $\operatorname{Min} P$ and $\operatorname{Max} P$ denote respectively the set of minimal elements and the set of maximal elements of P.

   (i) Prove that if X is a CTOD space then both $\operatorname{Min} X$ and $\operatorname{Max} X$ are non-empty provided X is. [Hint. Use the duality between **D** and **P**.]

   (ii) (For those at home with (ZL).) Give a direct proof of the claim in (i). [Hint. When proving that any chain C in X has an upper bound in X, use Exercise 10.9(ii) and the compactness of X to show that $\bigcap_{x \in C} \uparrow x \neq \varnothing$.]

10.11 Characterize the atoms (i) in $\mathcal{P}(X)$ for a Boolean space X, (ii) in $\mathcal{O}(X)$ for a CTOD space X.

10.12 A topological space X is called **extremally disconnected** if the closure of every open set in X is open. An ordered space X is called **extremally order-disconnected** if, for every open down-set U in X, the smallest closed down-set containing U is open.

   (i) Prove that a Boolean lattice B is complete if and only if the Boolean space $D(B)$ is extremally disconnected.

   (ii) Prove that a bounded distributive lattice L is complete if and only if the space $D(L)$ is extremally order-disconnected.

[Hint. Apply the duality. In (i), work with a Boolean space X and the Boolean lattice $\mathcal{P}(X)$. In (ii), apply Exercise 10.9(iii) to a CTOD space X and the distributive lattice $\mathcal{O}(X)$.]

10.13 Let X be a CTOD space and let $L = \mathcal{O}(X)$.

   (i) Prove that J is a principal ideal in L if and only if $\Phi(J)$ is a clopen down-set, where Φ is as in **10.24**.

   (ii) Prove that I is a prime ideal in L if and only if $\Phi(I) = X \setminus \uparrow x$ for some $x \in X$.

   (iii) Prove that I is a maximal ideal in L if and only if $\Phi(I) = X \setminus \{x\}$ for some maximal element x of X.

10.14 Let $L \in \mathbf{D}$. Use Exercise 10.13 to prove that every prime ideal in L is principal if and only if L is finite.

10.15 Let X be a CTOD space and let $L = \mathcal{O}(X)$. Describe the mutually inverse maps which establish a bijection between the lattice $\mathcal{F}(L)$ of filters of L and the lattice $\mathcal{F}$ of open up-sets in X.

10.16 A CTOD space X is called a p-**space** if $\uparrow U$ is open (and therefore clopen, by Exercise 10.9(ii)) for every clopen down-set U in X. Thus, by 10.23, $L = \mathcal{O}(X)$ is pseudocomplemented if and only if X is a p-space. Let X and Y be p-spaces; then a continuous order-preserving map $\varphi: Y \to X$ is called a p-**morphism**

if $\varphi(\text{Min } Y \cap {\downarrow} y) = \text{Min } X \cap {\downarrow}\varphi(y)$ for all $y \in Y$. Given pseudo-complemented lattices L and K, a map $f \colon L \to K$ is said to **preserve pseudocomplements** if $f(a^*) = f(a)^*$ for all $a \in L$.

   (i) Prove that, if X and Y are p-spaces and $\varphi \in \mathbf{P}(Y, X)$, then $E(\varphi) \colon \mathcal{O}(X) \to \mathcal{O}(Y)$ preserves pseudocomplements if and only if φ is a p-morphism.

   (ii) Show that if X is a p-space then $\text{Min } X$ is closed in X.

   (iii) Show that Examples X_2 and X_3 from Exercise 10.5 are not p-spaces while Examples X_4 and X_5 are.

   (iv) Let X be a CTOD space such that for every $x \in X$ there exists a unique element $m(x) \in \text{Min } X$ such that $m(x) \leqslant x$. Show that X is a p-space if and only if the map $m \colon X \to \text{Min } X$ is continuous.

10.17 Let B be a Boolean algebra. Show that the lattices $\text{Con } B$ of (Boolean) congruences, $\mathcal{I}(B)$ of ideals and $\mathcal{F}(B)$ of filters are isomorphic. [Hint. Use 10.27 and Exercise 7.10.]

Conclude that every (Boolean) congruence θ on a Boolean algebra is of the form θ_J for some ideal J in B (see Exercise 6.19).

10.18 This exercise and the next introduce the homset approach to duality referred to in 10.25. Let $L \in \mathbf{D}$ and $X \in \mathbf{P}$. The first task is to topologize $\mathbf{D}(L, \mathbf{2})$. Let $a \in L$ and $\varepsilon \in \{0, 1\}$ and let

$$U(a, \varepsilon) = \{\, f \in \mathbf{2}^L \mid f(a) = \varepsilon \,\}.$$

Let $\mathcal{T}$ be the topology on $\mathbf{2}^L$ which has the sets $U(a, \varepsilon)$, for $a \in L$ and $\varepsilon \in \{0, 1\}$, as a subbasis; thus U is open in $\mathbf{2}^L$ if and only if it is a union of sets each of which is a finite intersection of sets of the form $U(a, \varepsilon)$. We endow $\mathbf{D}(L, \mathbf{2})$ with the subspace topology; thus sets of the form

$$U(a, \varepsilon) \cap \mathbf{D}(L, \mathbf{2}) = \{\, f \in \mathbf{D}(L, \mathbf{2}) \mid f(a) = \varepsilon \,\}$$

constitute a subbasis for the topology on $\mathbf{D}(L, \mathbf{2})$. Show that the map $\varphi \colon \mathcal{I}_p(L) \to \mathbf{2}^L$, given by $\varphi(I) := f_I$ (see Exercise 9.2) is a homeomorphism onto $\mathbf{D}(L, \mathbf{2})$. [Hint. By Exercise 9.2, φ is one-to-one and maps onto $\mathbf{D}(L, \mathbf{2})$, so, by 10.7A, it remains to show that φ is continuous.]

10.19 Define maps $D' \colon \mathbf{D} \to \mathbf{P}$ and $E' \colon \mathbf{P} \to \mathbf{D}$ by

$$\begin{aligned} D' &\colon L \mapsto \mathbf{D}(L, \mathbf{2}) && (L \in \mathbf{D}), \\ E' &\colon X \mapsto \mathbf{P}(X, \mathbf{2}) && (X \in \mathbf{P}). \end{aligned}$$

Given $L, K \in \mathbf{D}$ and $g \in \mathbf{D}(L, K)$, define $D'(g): D'(K) \to D'(L)$ by $(D'(g))(f) = f \circ g$ for all $f \in D'(K) = \mathbf{D}(K, \mathbf{2})$ and, given $X, Y \in \mathbf{P}$ and $\psi \in \mathbf{P}(Y, X)$, define $E'(\psi): E'(X) \to E'(Y)$ by $(E'(\psi))(\varphi) = \varphi \circ \psi$ for all $\varphi \in E'(X) = \mathbf{P}(X, \mathbf{2})$.

(i) Show that $D'(g)$ is continuous and order-preserving and that $E'(\psi)$ is a $\{0, 1\}$-homomorphism.

(ii) For all $L \in \mathbf{D}$ and $X \in \mathbf{P}$ define $\eta_L: L \to E'D'(L)$ and $\varepsilon_X: X \to D'E'(X)$ to be the natural 'evaluation maps', that is, $(\eta_L(a))(f) := f(a)$ for all $a \in L$ and $f \in D'(K) = D(K, \mathbf{2})$ and, similarly, $(\varepsilon_X(x))(\varphi) := \varphi(x)$ for all $x \in X$ and all $\varphi \in E'(X) = \mathbf{P}(X, \mathbf{2})$. Show that η_L is a $\{0, 1\}$-homomorphism, that ε_X is continuous and order-preserving and that the diagrams below commute.

$$
\begin{array}{ccc}
L & \xrightarrow{\;f\;} & K \\
\eta_L \downarrow & & \downarrow \eta_K \\
E'D'(L) & \xrightarrow{E'D'(f)} & E'D'(K)
\end{array}
\qquad
\begin{array}{ccc}
Y & \xrightarrow{\;\varphi\;} & X \\
\varepsilon_Y \downarrow & & \downarrow \varepsilon_X \\
D'E'(Y) & \xrightarrow{D'E'(\varphi)} & D'E'(X)
\end{array}
$$

10.20 Let S be a set and equip $\mathbf{2}^S$ with the topology defined in the previous exercise. You may assume without proof that this topology is compact. (This is a consequence of a famous result known as Tychonoff's Theorem. Alternatively it can be obtained from Alexander's Subbasis Lemma.)

(i) Show that for all S the ordered space $\mathbf{2}^S$ belongs to $\mathbf{P}$.

(ii) Prove directly from the definition of the topology on $\mathbf{2}^L$ that $\mathbf{D}(L, \mathbf{2})$ is a closed subset of $\mathbf{2}^L$.

(iii) Let X be an ordered space. Show that $X \in \mathbf{P}$ if and only if there exists for some set S an order-homeomorphism φ from X onto a closed subspace of $\mathbf{2}^S$. [Hint. Use Exercise 10.19(i).]

10.21 Given a chain C define the **interval topology** on C to be the topology which has $\varnothing$, C and the sets of the form $\{\, x \in C \mid x < c \,\}$ and $\{\, x \in C \mid x > c \,\}$, for $c \in C$, as a subbasis.

(i) Prove that the interval topology on a chain C gives a Boolean space if and only if C is an algebraic lattice.

(ii) (For those who know about ordinals.) Show that the interval topology on an ordinal λ gives a Boolean space if and only if λ is not a limit ordinal.

11

Formal Concept Analysis

Hierarchies occur often both within mathematics and in the 'real' world. The theory of ordered sets and lattices provides a natural setting in which to discuss and analyse such hierarchies. In this chapter we take a brief excursion into **formal concept analysis** in order to get a feel for the potential of lattice theory in the analysis of hierarchies of concepts.

Contexts and their concepts

11.1 What is a concept? This would appear to be a question for philosophers rather than for mathematicians. Indeed, traditional philosophy's answer provides us with the basis for our formal definition. A concept is considered to be determined by its extent and its intent: the *extent* consists of all objects belonging to the concept (as the reader belongs to the concept 'living person') while the *intent* is the collection of all attributes shared by the objects (as all living persons share the attribute 'can breathe'). As it is often difficult to list all the objects belonging to a concept and usually impossible to list all its attributes, it is natural to work within a specific *context* in which the objects and attributes are fixed.

11.2 A context for the planets. The information presented in Table 11.1 gives a (somewhat limited) context for the planets of our solar system.

	size			distance from sun		moon	
	small	medium	large	near	far	yes	no
Mercury	×			×			×
Venus	×			×			×
Earth	×			×		×	
Mars	×			×		×	
Jupiter			×		×	×	
Saturn			×		×	×	
Uranus		×			×	×	
Neptune		×			×	×	
Pluto	×				×	×	

Table 11.1

The objects are the planets while the attributes are the seven indicated properties relating to size, distance from the sun and existence of

a moon. That the ith object possesses the jth attribute is indicated by
a $\times$ in the ij position of the table. A concept of this context will consist
of an ordered pair (A, B), where A (the *extent*) is a subset of the nine
planets and B (the *intent*) is a subset of the seven properties. To demand that the concept is determined by its extent and its intent means
that B should contain just those properties shared by all the planets
in A and, similarly, the planets in A should be precisely those sharing
all the properties in B. A simple procedure for finding a concept is as
follows: take an object, say the planet Earth, and let B be the set of
attributes which it possesses, in this case

$$B = \{\text{size-small, distance-near, moon-yes}\},$$

then let A be the set of all planets possessing all the attributes in B, in
this case

$$A = \{\text{Earth, Mars}\}.$$

Then (A, B) is a concept. More generally, we may begin with a set of
objects rather than a single object. Concepts may also be obtained via
a similar process commencing with a set of attributes.

It is usual to regard a concept (A_1, B_1) as being 'less general' than
a concept (A_2, B_2) if the extent A_1 of (A_1, B_1) is contained in the extent
A_2 of (A_2, B_2). Thus an order is defined on the set of concepts by

$$(A_1, B_1) \leqslant (A_2, B_2) \quad \Longleftrightarrow \quad A_1 \subseteq A_2.$$

The apparent asymmetry in this definition is illusory since $A_1 \subseteq A_2$ is
equivalent to $B_1 \supseteq B_2$ (see Lemma 11.4). The resulting ordered set
of concepts for our planetary context is the lattice given in Figure 11.1
later in the chapter. With respect to this order, the concept (A, B)
constructed above is the smallest concept whose extent contains the
planet Earth and is represented by the circle labelled EMa in Figure 11.1.
The significance of the labelling of the lattice will be explained when we
return to this example in 11.8.

We now wish to abstract the previous example. The resulting theory
has a broad range of applications outside mathematics, for example in
the social sciences, as well as having something useful to say within
lattice theory itself.

11.3 Concept Lattices.   A **context** is a triple (G, M, I) where G and
M are sets and $I \subseteq G \times M$. The elements of G and M are called **objects**
and **attributes** respectively. As usual, instead of writing $(g, m) \in I$ we
write gIm and say 'the object g has the attribute m'. (The letters G
and M come from the German: Gegenstände and Merkmale.)

For $A \subseteq G$ and $B \subseteq M$, define
$$A' = \{\, m \in M \mid (\forall g \in A)\, gIm \,\},$$
$$B' = \{\, g \in G \mid (\forall m \in B)\, gIm \,\};$$

so A' is the set of attributes common to all the objects in A and B' is the set of objects possessing the attributes in B. Then a **concept** of the context (G, M, I) is defined to be a pair (A, B) where $A \subseteq G$, $B \subseteq M$, $A' = B$ and $B' = A$. The **extent** of the concept (A, B) is A while its **intent** is B. Note that a subset A of G is the extent of some concept if and only if $A'' = A$ in which case the unique concept of which A is an extent is (A, A'). Of course, the corresponding statement applies to those subsets B of M which are the intent of some concept.

The set of all concepts of the context (G, M, I) is denoted by $\mathfrak{B}(G, M, I)$. (Again the choice of letter comes from the German: $\mathfrak{B}$ for Begriff.) For concepts (A_1, B_1) and (A_2, B_2) in $\mathfrak{B}(G, M, I)$ we write $(A_1, B_1) \leqslant (A_2, B_2)$, and say that (A_1, B_1) is a **subconcept** of (A_2, B_2), or that (A_2, B_2) is a **superconcept** of (A_1, B_1), if $A_1 \subseteq A_2$ (which by 11.4 is equivalent to $B_1 \supseteq B_2$). As we see in 11.5, $\langle \mathfrak{B}(G, M, I); \leqslant \rangle$ is a complete lattice; it is known as the **concept lattice** of the context (G, M, I).

There are two ways to go about proving the next lemma. One way is to work through the exercises on Galois connections (Exercises 11.2 to 11.4) and then note that the lemma is a particular case. Alternatively, the claims can be proved directly from the definitions of A' and B'. In either case, the details are left to the reader.

11.4 Lemma. *Assume that (G, M, I) is a context and let $A, A_j \subseteq G$ and $B, B_j \subseteq M$, for $j \in J$. Then*

    (i) $A \subseteq A''$,   (i)$'$ $B \subseteq B''$,

    (ii) $A_1 \subseteq A_2 \implies A_1' \supseteq A_2'$,   (ii)$'$ $B_1 \subseteq B_2 \implies B_1' \supseteq B_2'$,

    (iii) $A' = A'''$,   (iii)$'$ $B' = B'''$,

    (iv) $\left(\bigcup_{j \in J} A_j \right)' = \bigcap_{j \in J} A_j'$,   (iv)$'$ $\left(\bigcup_{j \in J} B_j \right)' = \bigcap_{j \in J} B_j'$.

The fundamental theorem

Before giving a range of examples of contexts and their concept lattices, we state and prove the characterization of concept lattices. Recall from 2.35 the definition of join-dense: a subset P of an ordered set Q is **join-dense** in Q if for every element $s \in Q$ there is a subset A of P such that $s = \bigvee_Q A$. Of course, **meet-dense** is the dual of join-dense.

11.5 The Fundamental Theorem on Concept Lattices. *Let (G, M, I) be a context. Then $\langle \mathfrak{B}(G, M, I); \leqslant \rangle$ is a complete lattice in which join and meet are given by*

$$\bigvee_{j \in J} (A_j, B_j) = ((\bigcup_{j \in J} A_j)'', \bigcap_{j \in J} B_j),$$

$$\bigwedge_{j \in J} (A_j, B_j) = (\bigcap_{j \in J} A_j, (\bigcup_{j \in J} B_j)'').$$

Conversely, if L is a complete lattice then L is isomorphic to $\mathfrak{B}(G, M, I)$ if and only if there are mappings $\gamma \colon G \to L$ and $\mu \colon M \to L$ such that $\gamma(G)$ is join-dense in L, $\mu(M)$ is meet-dense in L, and gIm is equivalent to $\gamma(g) \leqslant \mu(m)$ for each $g \in G$ and $m \in M$. In particular, L is isomorphic to $\mathfrak{B}(L, L, \leqslant)$ for every complete lattice L.

Proof. Although the fact that $\mathfrak{B}(G, M, I)$ is a complete lattice follows from general facts about Galois connections (see Exercises 11.2 to 11.4), we shall prove it directly. Clearly the relation $\leqslant$ is an order on $\mathfrak{B}(G, M, I)$. Let $(A_j, B_j) \in \mathfrak{B}(G, M, I)$ for all $j \in J$. By the preceding lemma, $\bigcap_{j \in J} A_j \subseteq (\bigcap_{j \in J} A_j)''$. Also $\bigcap_{j \in J} A_j \subseteq A_k$ for all $k \in J$, which implies that $(\bigcap_{j \in J} A_j)'' \subseteq A_k'' = A_k$ for all $k \in J$, whence $(\bigcap_{j \in J} A_j)'' \subseteq \bigcap_{k \in J} A_k = \bigcap_{j \in J} A_j$. Therefore $(\bigcap_{j \in J} A_j)'' = \bigcap_{j \in J} A_j$. Since, by 11.4(iv) ,

$$(\bigcap_{j \in J} A_j)' = (\bigcap_{j \in J} A_j'')' = (\bigcup_{j \in J} A_j')'' = (\bigcup_{j \in J} B_j)'',$$

it follows that

$$(\bigcap_{j \in J} A_j, (\bigcup_{j \in J} B_j)'') = (\bigcap_{j \in J} A_j, (\bigcap_{j \in J} A_j)') \in \mathfrak{B}(G, M, I).$$

Clearly $(\bigcap_{j \in J} A_j, (\bigcup_{j \in J} B_j)'')$ is a lower bound of $\{ (A_j, B_j) \mid j \in J \}$ in $\mathfrak{B}(G, M, I)$. If (A, B) is any lower bound of this set, then $A \subseteq A_j$ for all $j \in J$ and hence $A \subseteq \bigcap_{j \in J} A_j$, which gives $(A, B) \leqslant (\bigcap_{j \in J} A_j, (\bigcup_{j \in J} B_j)'')$. Thus

$$\bigwedge_{j \in J} (A_j, B_j) = (\bigcap_{j \in J} A_j, (\bigcup_{j \in J} B_j)'').$$

Since $(A_1, B_1) \geqslant (A_2, B_2)$ in $\mathfrak{B}(G, M, I)$ if and only if $B_1 \subseteq B_2$, replacing A by B virtually everywhere in the proof above shows that

$$\bigvee_{j \in J} (A_j, B_j) = ((\bigcup_{j \in J} A_j)'', \bigcap_{j \in J} B_j).$$

Now let φ be an order-isomorphism from $\mathfrak{B}(G, M, I)$ onto a complete lattice L. Define $\gamma \colon G \to L$ and $\mu \colon M \to L$ by

$$\gamma(g) := \varphi((\{g\}'', \{g\}')) \quad \text{and} \quad \mu(m) := \varphi((\{m\}', \{m\}''))$$

for all $g \in G$ and $m \in M$. For all $(A, B) \in \mathfrak{B}(G, M, I)$,

$$\bigvee \gamma(A) = \bigvee_{g \in A} \gamma(g)$$

$$= \bigvee_{g \in A} (\{g\}'', \{g\}')$$

$$= ((\bigcup_{g \in A} \{g\}'')'', \bigcap_{g \in A} \{g\}').$$

But, by 11.4(iv), for all $A \subseteq G$ we have

$$\bigcap_{g \in A} \{g\}' = (\bigcup_{g \in A} \{g\})' = A' = B.$$

Since (A, B) and $\bigvee \gamma(A)$ are elements of $\mathfrak{B}(G, M, I)$ with the same second coordinate, $\bigvee \gamma(A) = (A, B)$. Similarly, for all $(A, B) \in \mathfrak{B}(G, M, I)$ we find $\bigwedge \mu(B) = (A, B)$. Consequently $\gamma(G)$ is join-dense and $\mu(M)$ is meet-dense in L. Furthermore, for all $g \in G$ and $m \in M$,

$$gIm \iff g \in \{m\}'$$
$$\iff \{g\} \subseteq \{m\}'$$
$$\iff \{g\}'' \subseteq \{m\}''' = \{m\}'$$
$$\iff \gamma(g) \leqslant \mu(m).$$

(The forward direction of the third equivalence above uses 11.4(ii), (ii)′ and (iii)′ while the backward implication uses 11.4(i).)

Conversely, assume that L is a complete lattice and that $\gamma \colon G \to L$ and $\mu \colon M \to L$ are mappings such that $\gamma(G)$ is join-dense and $\mu(M)$ is meet-dense in L and gIm holds if and only if $\gamma(g) \leqslant \mu(m)$. Before defining the order-isomorphism between $\mathfrak{B}(G, M, I)$ and L we prove that, for all $A \subseteq G, B \subseteq M, g \in A$ and $m \in M$,

$$m \in A' \iff \bigvee \gamma(A) \leqslant \mu(m) \quad \text{and} \quad g \in B' \iff \gamma(g) \leqslant \bigwedge \mu(B).$$

By duality, it suffices to prove the left-hand equivalence. For all $m \in M$ we have

$$m \in A' \implies (\forall g \in A) \, gIm$$
$$\implies (\forall g \in A) \, \gamma(g) \leqslant \mu(m) \qquad \text{(by assumption)}$$
$$\implies \mu(m) \text{ is an upper bound of } \gamma(A)$$
$$\implies \bigvee \gamma(A) \leqslant \mu(m).$$

Conversely, we have

$$\bigvee \gamma(A) \leqslant \mu(m) \implies (\forall g \in A)\, \gamma(g) \leqslant \mu(m) \ \ (\text{as } \gamma(g) \leqslant \bigvee \gamma(A) \leqslant \mu(m))$$
$$\implies (\forall g \in A)\, gIm \qquad\qquad (\text{by assumption})$$
$$\implies m \in A'.$$

We shall now define a pair of mutually inverse order-preserving maps $\varphi \colon \mathfrak{B}(G, M, I) \to L$ and $\psi \colon L \to \mathfrak{B}(G, M, I)$. Define φ by $\varphi((A, B)) = \bigvee \gamma(A)$ for all $(A, B) \in \mathfrak{B}(G, M, I)$. Since

$$(A_1, B_1) \leqslant (A_2, B_2) \implies A_1 \subseteq A_2$$
$$\implies \gamma(A_1) \subseteq \gamma(A_2)$$
$$\implies \bigvee \gamma(A_1) \leqslant \bigvee \gamma(A_2) \quad (\text{by } 2.8(\mathrm{v}))$$

the map φ is order-preserving. Define ψ by $\psi(x) = (A_x, B_x)$ where

$$A_x := \{\, g \in G \mid \gamma(g) \leqslant x \,\} \quad \text{and} \quad B_x := \{\, m \in M \mid x \leqslant \mu(m) \,\},$$

for each $x \in L$. For all $m \in M$ we have

$$m \in A_x' \iff \bigvee \gamma(A_x) \leqslant \mu(m) \qquad (\text{from above})$$
$$\iff x \leqslant \mu(m) \qquad\qquad (\text{since } \gamma(G) \text{ is join-dense in } L)$$
$$\iff m \in B_x.$$

Consequently $A_x' = B_x$ and dually (since $\mu(M)$ is meet-dense in L) we find $B_x' = A_x$. Thus (A_x, B_x) is a concept of (G, M, I), whence ψ is well defined. Since

$$x \leqslant y \implies A_x \subseteq A_y \implies (A_x, B_x) \leqslant (A_y, B_y),$$

ψ is order-preserving. Clearly

$$\varphi(\psi(x)) = \varphi((A_x, B_x)) = \bigvee \gamma(A_x) = x,$$

since $\gamma(G)$ is join-dense in L. Now let $(A, B) \in \mathfrak{B}(G, M, I)$; we shall prove that $\psi(\varphi((A, B))) = (A, B)$. Let $x := \varphi((A, B)) = \bigvee \gamma(A)$; thus we wish to show that $(A_x, B_x) = (A, B)$, for which it suffices to check that $A_x = A$. For all $g \in A$ we have $\gamma(g) \leqslant \bigvee \gamma(A) = x$ and thus $g \in A_x$. Hence $A \subseteq A_x$. In the other direction,

$$g \in A_x \implies \gamma(g) \leqslant x = \bigvee \gamma(A)$$
$$\implies (\forall m \in A')\, \gamma(g) \leqslant \mu(m) \quad (\text{since } (\forall m \in A')\, \bigvee \gamma(A) \leqslant \mu(m))$$
$$\implies (\forall m \in A')\, gIm \qquad\qquad (\text{by assumption})$$
$$\implies g \in A'' = A \qquad\qquad (\text{since } (A, B) \text{ is a concept}).$$

Thus $A_x \subseteq A$ and hence $A_x = A$, as required.

Hence φ and ψ are order-preserving and mutually inverse, whence $\mathfrak{B}(G, M, I)$ is order-isomorphic to L by 1.13(4).

Given a complete lattice L, we can choose $G = M = L, I = \leqslant$, and $\gamma = \mu = \mathrm{id}_L$ (the identity map on L). The conditions of the theorem are clearly satisfied and hence $L \cong \mathfrak{B}(L, L, \leqslant)$. ∎

11.6 Remarks. It follows from 11.4 that, given a context (G, M, I), the map $A \mapsto A''$ defines a closure operator on the set G of objects and similarly $B \mapsto B''$ yields a closure operator on the set M of attributes. By 2.21, the corresponding topped $\bigcap$-structures

$$\mathfrak{B}_G := \{ A \subseteq G \mid A'' = A \} \quad \text{and} \quad \mathfrak{B}_M := \{ B \subseteq M \mid B'' = B \}$$

form complete lattices when ordered by inclusion. It is an extremely easy exercise to see that $(A, B) \mapsto A$ gives an order-isomorphism between $\mathfrak{B}(G, M, I)$ and $\mathfrak{B}_G$ while $(A, B) \mapsto B$ gives an order-isomorphism between $\mathfrak{B}(G, M, I)$ and $\mathfrak{B}_M^\partial$. We choose to work here with $\mathfrak{B}(G, M, I)$ rather than $\mathfrak{B}_G$ or $\mathfrak{B}_M$ since we want both the extent and the intent of each concept to be visible.

From theory to practice

We begin with some inbred examples from lattice theory itself.

11.7 Lattice-theoretic examples.

(1) If L is a complete lattice, then the Fundamental Theorem yields $L \cong \mathfrak{B}(L, L, \leqslant)$. In fact,

$$\mathfrak{B}(L, L, \leqslant) = \{ (\downarrow x, \uparrow x) \mid x \in L \}$$

and the isomorphism is given by $x \mapsto (\downarrow x, \uparrow x)$.

(2) For a given complete lattice L, the context $(L, L, \leqslant)$ is not the only one whose concept lattice is isomorphic to L, nor is it necessarily the most natural one. Consider the complete lattice $\wp(X)$ for some set X. While the concept lattice of the context $(\wp(X), \wp(X), \subseteq)$ is isomorphic to $\wp(X)$, both of the following concept lattices have the property that $\mathfrak{B}_G$ (as defined in 11.6) actually equals $\wp(X)$.

  (i) $\mathfrak{B}(X, X, \neq) = \{ (A, X \smallsetminus A) \mid A \subseteq X \} \cong \mathfrak{B}_G = \wp(X)$.

  (ii) $\mathfrak{B}(X, \wp(X), \in) = \{ (A, \{ B \in \wp(X) \mid A \subseteq B \}) \mid A \subseteq X \} \cong \mathfrak{B}_G = \wp(X)$.

(3) In contrast to (2)(i) above, we have

$$\mathfrak{B}(X, X, =) = \{ (\{x\}, \{x\}) \mid x \in X \} \cup \{ (\varnothing, X), (X, \varnothing) \}.$$

Hence, if $|X| = n$, then $\mathfrak{B}(X, X, =) \cong \mathfrak{B}_G \cong \mathbf{M}_n$.

(4) This example generalizes (1). Let $(P; \leqslant)$ be an ordered set and consider the context $(P, P, \leqslant)$. For $A \subseteq P$ (regarded as a set of objects) we have $A'' = A^{u\ell}$ and hence $\mathfrak{B}_G$ is the Dedekind–MacNeille completion of P (see 2.32). Thus

$$\mathfrak{B}(P, P, \leqslant) = \{\, (A, A^u) \mid A \in \mathbf{DM}(P)\,\} \cong \mathfrak{B}_G = \mathbf{DM}(P).$$

If L is a complete lattice, then $\mathbf{DM}(L) \cong L$ (by 2.37(iii)), and consequently (1) is a special case of this example.

(5) Let L be a lattice with no infinite chains, in particular any finite lattice. Then by 2.28(iii) (or just 2.12 if L is finite) L is complete and by 7.10 the subsets $\mathcal{J}(L)$ of join-irreducible elements and $\mathcal{M}(L)$ of meet-irreducible elements are join-dense and meet-dense in L, respectively. Thus the Fundamental Theorem yields $L \cong \mathfrak{B}(\mathcal{J}(L), \mathcal{M}(L), \leqslant)$ with the isomorphism given by

$$x \mapsto (\downarrow x \cap \mathcal{J}(L), \uparrow x \cap \mathcal{M}(L)).$$

We turn now to non-mathematical examples.

11.8 Returning to the planets. The concept lattice of the planetary context given in Table 11.1 and considered earlier in 11.2 is presented in Figure 11.1. The labels indicate the mappings γ and μ of the Fundamental Theorem, that is, the circle labelled g represents the concept $(\{g\}'', \{g\}')$ for all $g \in G$ and similarly the circle labelled m represents the concept $(\{m\}', \{m\}'')$ for each $m \in M$. The map $\psi \colon L \mapsto \mathfrak{B}(G, M, I)$ defined in the proof of the Fundamental Theorem allows us to read off the extent and intent of each concept from the labelling: if x is an element of the lattice, then the corresponding concept is

$$(\{\, g \in G \mid \gamma(g) \leqslant x \,\}, \{\, m \in M \mid x \leqslant \mu(m) \,\}).$$

For example, the element in the middle of the diagram corresponds to the concept

$$(\{\text{Earth, Mars, Pluto}\}, \{\text{size-small, moon-yes}\}).$$

The concept lattice provides a basic analysis of a context: it yields an appropriate classification of the objects and at the same time indicates the implications between the attributes. For the concept lattice of a context to be of practical use, we must be able to *determine* which pairs (A, B), with $A \subseteq G$ and $B \subseteq M$, are concepts of the context, and we must then be able to *describe* the resulting lattice of concepts.

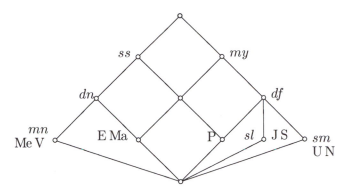

Figure 11.1

11.9 The determination problem. A simple-minded and extremely inefficient way of determining all the concepts of a context (G, M, I) would be to form (A'', A') for all $A \subseteq G$ (or (B', B'') for all $B \subseteq M$). Here is an efficient alternative: for $A \subseteq G$ and $B \subseteq M$,

$$B' = \bigcap_{m \in B} \{m\}' \quad \text{and} \quad A' = \bigcap_{g \in A} \{g\}'.$$

In particular, if (A, B) is a concept of the context (G, M, I), then

$$A = \bigcap_{m \in B} \{m\}' \quad \text{and} \quad B = \bigcap_{g \in A} \{g\}'.$$

For example, in the context for the planets considered in 11.2 and 11.8, we first determine the extents

$$\{ss\}', \{sm\}', \{sl\}', \{dn\}', \{df\}', \{my\}' \text{ and } \{mn\}'$$

and then obtain all other extents by forming intersections. The intent corresponding to each extent is then easily calculated. Alternatively, we first determine all the intents and then the corresponding extents.

11.10 A psychometric test. The context given in Table 11.2 shows the characteristics attributed by a patient to his relatives as well as his own ideal. The concept lattice (see Figure 11.2) is intended in this case to aid therapists in their analysis of the patient's responses. Note, for example, that the concept

$$(\{Se, F, M, Si\}, \{an, d, r\})$$

indicates that the patient characterizes his immediate family (Self, Father, Mother and Sister) as being anxious, difficult and reserved. In this case, calculating $\{m\}'$ for each of the fourteen choices for the attribute

	Self	My Ideal	Father	Mother	Sister	Brother-in-Law
vulnerable	×	×	×	×	×	
reserved	×		×	×	×	
self-confident	×	×				×
dutiful		×	×	×	×	×
happy	×	×	×	×	×	×
difficult	×		×	×	×	
attentive	×	×	×		×	×
easily offended			×	×		
not hot-tempered	×	×	×	×	×	
anxious	×		×	×	×	
talkative						×
superficial			×	×		×
sensitive	×	×	×	×	×	
ambitious	×	×	×	×	×	×

Table 11.2

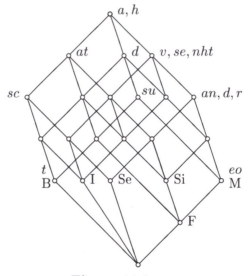

Figure 11.2

$m \in M$ leads to eight distinct extents while the six objects $g \in G$ give rise to six distinct intents of the form $\{g\}'$. With these in hand it is easy to find the nine remaining elements of the lattice.

11.11 The description problem. It must go almost without saying at this stage that the most natural and informative description of the concept lattice $\mathfrak{B}(G, M, I)$ is a well-drawn diagram labelled by (names for) the elements of G and M. If the number of concepts is small then the task may be done by hand. As the size of the context and its lattice of concepts grows, the need for appropriate computer software becomes more apparent. The theory of 'readable diagrams' and computer-based implementations of such a theory is in its infancy. (See R. Wille [55] and the references therein.) The development of this theory and related software is sure to be one of the growth industries amongst the next generation of lattice-and-order theorists.

Exercises

11.1 Draw and label the concept lattice for each of the contexts below. In each case the incidence relation I is given in Table 11.3 with G listed vertically and M horizontally.

(1) Large German cities.

$$G = \{\text{Hamburg, München, Köln}\},$$
$$M = \{\geqslant \textbf{1.5 million}, \geqslant \textbf{1.25 million}, \geqslant \textbf{1.0 million}\}.$$

(2) Triangular shapes.

$$G = \{\text{Setsquare, Give Way Sign, Delta}\},$$
$$M = \{\text{rightangled, isosceles, equilateral}\}.$$

(3) Primary colour decomposition.

$$G = \{\text{Orange, Green, Violet}\},$$
$$M = \{\text{blue, red, yellow}\}.$$

(4) Temperatures.

$$G = \{\text{Cold, Tepid, Warm}\},$$
$$M = \{\leqslant 10°\text{C}, \leqslant 20°\text{C}, \geqslant 10°\text{C}, \geqslant 20°\text{C}\}.$$

(5) Watercourses.

$$G = \{\text{Channel, Brook, Stream, River}\},$$
$$M = \{\text{very small, small, large, very large}\}.$$

(6) Family.

$$G = \{\text{Father, Mother, Son, Daughter}\},$$
$$M = \{\text{old, young, male, female}\}.$$

	≥1.5m	≥1.25m	≥1.0m
H	×	×	×
M		×	×
K			×

Context (1)

	r	i	e
S	×		
G		×	×
D		×	

Context (2)

	b	r	y
O		×	×
G	×		×
V	×	×	

Context (3)

	≤10°	≤20°	≥10°	≥20°
C	×	×		
T		×	×	
W			×	×

Context (4)

	vs	s	l	vl
C	×	×		
B		×		
S			×	
R			×	×

Context (5)

	o	y	m	f
F	×		×	
M	×			×
S		×	×	
D		×		×

Context (6)

Table 11.3

11.2 Let P be an ordered set. Then a map $c\colon P \to P$ is called a **closure operator** (on P) if for all $x, y \in P$

(a) $x \leqslant c(x)$,

(b) $x \leqslant y \Longrightarrow c(x) \leqslant c(y)$,

(c) $c(c(x)) = c(x)$.

An element $x \in P$ is called **closed** if $c(x) = x$. The set of all closed elements of P is denoted by P_c.

(i) Show that $P_c = \{ c(x) \mid x \in P \}$ and that P_c contains the top element of P when it exists.

(ii) Prove that if P is a complete lattice then P_c is a complete lattice such that, for every subset S of P_c,

$$\bigwedge_{P_c} S = \bigwedge_{P} S \text{ and } \bigvee_{P_c} S = c\left(\bigvee_{P} S\right).$$

[Hint. Use Theorem 2.21.]

11.3 Let P and Q be ordered sets. A pair $(\blacktriangleright, \blacktriangleleft)$ of maps $\blacktriangleright\colon P \to Q$ and $\blacktriangleleft\colon Q \to P$ (which we refer to as *right* and *left* respectively) is called a **Galois connection** between P and Q if

(a) $p \leqslant q^{\blacktriangleleft} \Leftrightarrow q \leqslant p^{\blacktriangleright}$ for all $p \in P$ and $q \in Q$.

(i) Show that $(^\blacktriangleright, {}^\blacktriangleleft)$ is a Galois connection between P and Q if and only if

    (b) $^\blacktriangleright$ and $^\blacktriangleleft$ are both order-reversing, and

    (c) $p \leqslant p^{\blacktriangleright\blacktriangleleft}$ for all $p \in P$ and $q \leqslant q^{\blacktriangleleft\blacktriangleright}$ for all $q \in Q$.

(ii) Let $(^\blacktriangleright, {}^\blacktriangleleft)$ be a Galois connection. Show that

    (d) $p^{\blacktriangleright\blacktriangleleft\blacktriangleright} = p^{\blacktriangleright}$ and $q^{\blacktriangleleft\blacktriangleright\blacktriangleleft} = q^{\blacktriangleleft}$ for all $p \in P$ and $q \in Q$.

(iii) Let $(^\blacktriangleright, {}^\blacktriangleleft)$ be a Galois connection. Show that

$$c \colon P \to P \colon p \mapsto p^{\blacktriangleright\blacktriangleleft} \quad \text{and} \quad k \colon Q \to Q \colon q \mapsto q^{\blacktriangleleft\blacktriangleright}$$

are closure operators on P and Q, respectively. Prove that $^\blacktriangleright \colon P_c \to Q_k$ and $^\blacktriangleleft \colon Q_k \to P_c$ are mutually inverse, (well-defined) dual order-isomorphisms between the sets

$$P_c = \{\, p \in P \mid p^{\blacktriangleright\blacktriangleleft} = p \,\} \quad \text{and} \quad Q_k = \{\, q \in Q \mid q^{\blacktriangleleft\blacktriangleright} = q \,\}$$

of closed elements of P and Q, respectively. (Hence, by Exercise 11.2, if P and Q are complete lattices then P_c and Q_k are dually order-isomorphic complete lattices.)

(iv) Given a Galois connection $(^\blacktriangleright, {}^\blacktriangleleft)$ and subsets A of P and B of Q, we define

$$A^{\blacktriangleright} := \{\, p^{\blacktriangleright} \mid p \in A \,\} \quad \text{and} \quad B^{\blacktriangleleft} := \{\, q^{\blacktriangleleft} \mid q \in B \,\}.$$

Show that if $\bigvee_P A$ exists in P, then $\bigwedge_Q A^{\blacktriangleright}$ exists in Q and $\bigwedge_Q A^{\blacktriangleright} = (\bigvee_P A)^{\blacktriangleright}$, and similarly, if $\bigvee_Q B$ exists in Q, then $\bigwedge_P B^{\blacktriangleleft}$ exist in P and $\bigwedge_P B^{\blacktriangleleft} = (\bigvee_Q B)^{\blacktriangleleft}$.

11.4 Assume that (G, M, I) is a context. Show that the pair of maps

$$' \colon \wp(G) \to \wp(M) \quad \text{and} \quad ' \colon \wp(M) \to \wp(G),$$

defined in 11.3, yields a Galois connection between $\wp(G)$ and $\wp(M)$. Hence use Exercise 11.3 to verify Lemma 11.4.

11.5 Let G and M be finite sets.

(i) Assume that $\langle G; \leqslant \rangle$ and $\langle M; \leqslant \rangle$ are chains and $I \subseteq G \times M$ is a down-set of the product. Show that $\mathfrak{B}(G, M, I)$ is a chain.

(ii) Assume that (G, M, I) is a context such that $\mathfrak{B}(G, M, I)$ is a chain. Show that G and M may be linearly ordered such that I becomes a down-set of $G \times M$. [Hint. choose orders on G and M such that $\langle G; \leqslant \rangle$ and $\langle M; \leqslant \rangle$ are chains and $\gamma \colon G \to$

$\mathfrak{B}(G, M, I)$ is order-preserving while $\mu\colon M \to \mathfrak{B}(G, M, I)$ is order-reversing.]

11.6 Let P be an ordered set and consider the context $(P, P, \not\geq)$. Prove that $\mathfrak{B}(P, P, \not\geq)$ is isomorphic to the (distributive) lattice $\mathcal{O}(P)$ of down-sets of P.

11.7 Find all concepts of the planetary context discussed in 11.2 and 11.8 and verify that the lattice given in Figure 11.1 is correct.

11.8 Repeat Exercise 11.7 for the context of the psychometric test given in 11.10.

11.9 Consider the task-information context given in Table 11.4 which comes from the German national division of health and welfare. The context shows which personal information is required by law for the different tasks of a local medical subdivision. The tasks (objects) and information (attributes) are as follows:
TASKS

1. confirmation of requests for preventative care;
2. calculation of insurance benefits;
3. confirmation of incapacity to work;
4. confirmation of correct diagnosis;
5. work preliminary to rehabilitation;
6. verification of sickness;
7. advice to clients;
8. advice on general preventatives;
9. carrying on the statistics of the medical service.

INFORMATION

a. name and address of client;
b. career history;
c. kind of membership;
d. name of responsible agency;
e. family medical history;
f. vocational education;
g. number of certificates.

Find the 13 concepts of this context then draw and label the resulting concept lattice.

11.10 The concept lattice of a task-information context (G, M, I) like the one in the previous example provides a natural hierarchical classification of the tasks and indicates the dependencies between the

	a	b	c	d	e	f	g
1	×	×	×	×	×	×	
2	×	×	×	×	×	×	
3	×	×	×	×	×	×	
4	×	×	×	×			
5	×	×	×	×	×	×	
6	×		×	×		×	
7	×	×	×		×	×	×
8							
9			×				×

Table 11.4

information. In some situations it may be more important to know which information is not required for a particular task; for example, if we wish to restrict access to personal or classified information. Then it is more appropriate to work with the **complementary context** $(G, \overline{M}, \overline{I})$ where

$$\overline{M} := \{\, \overline{m} \mid m \in M \,\} \text{ and } g\overline{I}\overline{m} \Longleftrightarrow \neg(gIm).$$

(Here $\overline{m}$ is a new symbol to be thought of as 'not m'.) Consider the information-task context in the previous exercise. Find the 11 concepts of the complementary context then draw and label the resulting lattice $\mathfrak{B}(G, \overline{M}, \overline{I})$.

11.11 The **horizontal sum** of two bounded ordered sets P and Q is obtained from their disjoint union $P \,\dot\cup\, Q$ by identifying the bottoms of the two ordered sets and also identifying the tops. The **vertical sum** of P and Q is obtained from the linear sum $P \oplus Q$ by identifying the top of P with the bottom of Q.

Let (G_1, M_1, I_1) and (G_2, M_2, I_2) be contexts with $G_1 \cap G_2 = M_1 \cap M_2 = \emptyset$ and $G_i' = M_i' = \emptyset$ for $i = 1, 2$ and let $L_i := \mathfrak{B}(G_i, M_i, I_i)$ for $i = 1, 2$. Prove the following claims.

(i) $\mathfrak{B}(G_1 \,\dot\cup\, G_2, M_1 \,\dot\cup\, M_2, I_1 \,\dot\cup\, I_2)$ is isomorphic to the horizontal sum of L_1 and L_2.

(ii) $\mathfrak{B}(G_1 \,\dot\cup\, G_2, M_1 \,\dot\cup\, M_2, I_1 \,\dot\cup\, I_2 \,\dot\cup\, (G_1 \times M_2))$ is isomorphic to the vertical sum of L_1 and L_2.

(iii) $\mathfrak{B}(G_1 \,\dot\cup\, G_1, M_1 \,\dot\cup\, M_2, I_1 \,\dot\cup\, I_2 \,\dot\cup\, (G_1 \times M_2) \,\dot\cup\, (G_2 \times M_1))$ is isomorphic to the direct product of L_1 and L_2.

11.12 Use the previous exercise to find the concept lattice for each of the contexts given in Table 11.5.

In each case draw the lattice. (You need not label it.)

	a	b	c	d	e	f
A		×	×	×	×	×
B			×	×	×	×
C		×		×	×	×
D	×	×	×		×	×
E	×	×	×			×
F	×	×	×			

Context (i)

	a	b	c	d	e	f
A		×	×	×	×	×
B	×		×	×	×	×
C				×	×	×
D		×			×	×
E				×		×
F					×	

Context (ii)

	a	b	c	d	e	f	g	h
A		×	×	×				
B	×		×	×				
C				×				
D	×		×					
E						×	×	×
F							×	×
G						×		×
H								

Context (iii)

Table 11.5

Appendix: Further Reading

Background references for related areas of mathematics.

[1] T. S. Blyth, *Introduction to categories*, Longman, London, New York, 1982.

[2] N. Bourbaki, *Éléments de mathématiques: Topologie générale*, Hermann, Paris. (English translation published by Addison-Wesley, Mass.)

[3] S. Burris and H. P. Sankappanavar, *A course in universal algebra*, Springer Verlag, New York, Heidelberg, Berlin, 1981.

[4] C. C. Chang and K. J. Keisler, *Model theory*, North-Holland, Amsterdam, 1973.

[5] P. M. Cohn, *Algebra*, vols. I and II, Wiley, London, 1974 and 1976.

[6] P. M. Cohn, *Universal algebra*, Harper and Row, New York, 1965.

[7] J. Dugundji, *Topology*, Allyn and Bacon, Boston, 1966.

[8] H. B. Enderton, *Elements of set theory*, Academic Press, New York, 1977.

[9] J. B. Fraleigh, *A first course in abstract algebra*, 4th edition, Addison-Wesley, 1989.

[10] G. Grätzer. *Universal algebra*, 2nd edition, Springer Verlag, Berlin, Heidelberg, New York, 1979.

[11] A. G. Hamilton, *Numbers, sets and axioms*, Cambridge University Press, Cambridge, 1982.

[12] A. G. Hamilton, *Logic for mathematicians*, 2nd edition, Cambridge University Press, Cambridge, 1988.

[13] J. Kelley, *General topology*, Van Nostrand, Princeton N. J., 1955.

[14] S. Maclane, *Categories for the working mathematician*, Springer Verlag, Berlin, Heidelberg, New York, 1971.

[15] S. Maclane and G. Birkhoff, *Algebra*, Macmillan, New York, London, 1967.

[16] J. S. Rose, *A course on group theory*, Cambridge University Press, Cambridge, 1978.

[17] W. A. Sutherland, *An introduction to metric and topological spaces*, Oxford University Press, 1975.

General references on the theory of ordered sets and lattices. Few books have been written on lattices and ordered sets. The successive editions of G. Birkhoff's *Lattice theory* (1940, 1948 and 1967) are pioneering classics. In the 1960s George Grätzer put forward a proposal for a survey of the whole of lattice theory in depth, of which his textbook [23] on distributive lattices was originally intended as the first part. The rapid development of lattice theory in the following decade quickly made Grätzer's original objective quite impossible to attempt and [24], which appeared in 1978, had more restricted aims. More recently, Ralph McKenzie, George McNulty and Walter Taylor have embarked on a four volume work on the theory of algebras, in which lattices play a central role. This, of which [26] forms the first volume, will undoubtedly become a standard advanced reference.

M. Erné's enchantingly illustrated text [21] covers much of the same ground as our Chapters 1 and 2, and provides links with set theory. The Compendium, [22], and [25] are essential reading, at an advanced level, for those wishing to pursue further the order-theoretic background to domain theory and the connections between order, lattices and topology.

Proceedings of conferences supplement the textbook literature. In particular we draw attention to [28]. This contains a range of interesting articles which give access through their bibliographies to applications of order theory (including those in the social sciences) which we do not have space to reference individually. In addition [28] contains a comprehensive bibliography for ordered sets, up to 1981. Lattices and ordered sets are also served by two specialist journals, *Algebra Universalis* and *Order*, which were launched in 1971 and 1984, respectively.

[18] R. Balbes and Ph. Dwinger, *Distributive lattices*, University of Missouri Press, Missouri, 1974.

[19] G. Birkhoff, *Lattice theory*, 3rd edition, Coll. Publ., XXV, American Mathematical Society, Providence, R. I., 1967.

[20] P. Crawley and R.P. Dilworth, *Algebraic theory of lattices*, Prentice-Hall, N. J., 1973.

[21] M. Erné, *Einführung in die Ordnungstheorie*, B. I. Verlag, Mannheim, Wein, Zürich, 1982.

[22] G. Gierz, K. H. Hofmann, K. Keimel, J. D. Lawson, M. Mislove and D. S. Scott, *A compendium of continuous lattices*, Springer Verlag, Berlin, Heidelberg, New York, 1980.

[23] G. Grätzer, *Lattice theory: first concepts and distributive lattices*, W. H. Freeman and Co., San Francisco, 1971.

[24] G. Grätzer, *General lattice theory*, Birkhäuser, Basel, 1978.

[25] P. T. Johnstone, *Stone spaces*, Cambridge University Press, Cambridge, 1982.

[26] R. N. McKenzie, G. F. McNulty and W. F. Taylor, *Algebras, lattices and varieties*, vol. I, Wadsworth and Brooks, Monterey, California, 1987.

[27] M. Pouzet and D. Richard (eds.), *Orders: description and roles* (*Annals of Discrete Mathematics* **23**), North-Holland, Amsterdam, 1984.

[28] I. Rival (ed.), *Ordered sets*, NATO ASI Series **83**, Reidel, Dordrecht, 1982.

[29] I. Rival (ed.), *Graphs and order*, NATO ASI Series **147**, Reidel, Dordrecht, 1985.

[30] I. Rival (ed.), *Algorithms and order*, NATO ASI Series **255**, Reidel, Dordrecht, 1989.

[31] I. Rival, Order: a theory with a view, in *Klassifikation und Ordnung*, INDEKS Verlag, Frankfurt, 1989.

We do not attempt to attribute all the theorems in the book, but merely give references for a few of the major landmarks, as well as selected suggestions for background and further reading on specialized topics.

Chapters 1 and 2.

[32] B. Banaschewski and G. Bruns, Injective hulls in the category of distributive lattices, *J. Reine Angew. Math.* **232** (1968), 102–109. [The source for Theorem 2.36.]

[33] H. M. MacNeille, Partially ordered sets, *Trans. Amer. Math. Soc.* **42** (1937), 90–96. [MacNeille's orginal paper on completion.]

[34] H. Tverberg, On Dilworth's decomposition theorem for partially ordered sets, *J. Combinatorial Theory* **3** (1967), 305–6. [See Exercise 1.23.]

Chapters 3 and 4. Our treatment of CPOs, domains and fixpoint theory has been much influenced by unpublished notes by D.S. Scott, S. Abramsky and A.W. Roscoe, as well as by the sources below. Our treatment of Zorn's Lemma in Chapter 4 is complemented by that in [8] and [11].

[35] N. Bourbaki, Sur la théoreme de Zorn, *Arch. Math. (Basel)* **2** (1949/50), 434–437. [CPO Fixpoint Theorems II and III.]

[36] J. D. Lawson, The versatile continuous order, *Lecture Notes in Computer Science* **298**, ed. M. Main *et al.*, Springer Verlag, 1987, pp. 134–160.

[37] J. Loeckx and K. Sieber, *The foundations of program verification*, 2nd edition, Wiley–Teubner, Stuttgart, 1987.

[38] G. Markowsky, Chain-complete posets and directed sets with applications, *Algebra Univ.* **6** (1976), 53–68.

[39] D. A. Schmidt, *Denotational semantics*, Allyn and Bacon, Boston, 1986.

[40] D. S. Scott, Towards a mathematical theory of computation, *Proc. 4th Ann. Princeton Conf. on Inform. Sci. and Systems*, 1970, 169–176.

[41] D. S. Scott, Domains for denotational semantics, *Lecture Notes in Computer Science* **140**, ed. M. Nielsen and E. T. Schmidt, Springer Verlag, 1982, pp. 577–613.

[42] G. Winskel and K. G. Larsen, Using information systems to solve domain equations recursively, *Lecture Notes in Computer Science* **173**, ed. G. Kahn and G. D. Plotkin, Springer Verlag, 1984, pp. 109–130.

Chapters 5 and 6.

[43] G. Birkhoff, Applications of lattice algebra, *Proc. Camb. Phil. Soc.* **30** (1934), 115–122. [Theorem 6.10(ii).]

[44] R. Dedekind, Über die von drei Moduln erzeugte Dualgruppe, *Math. Ann.* **53** (1900), 371–403. [Modularity of $\mathcal{N}$–Sub G (6.6(5)) and Theorem 6.10(i), implicitly.]

Chapters 7 and 9.

[45] J. L. Bell and A. B. Slomson,*Models and ultraproducts*, North-Holland, Amsterdam, 1971.

[46] G. Boole, *An investigation into the laws of thought*, London, 1854 (reprinted by Open Court Publishing Co., Chicago, 1940).

[47] P. Halmos, *Lectures on Boolean algebras*, Van Nostrand, Princeton, N. J., 1963.

[48] A. S. Tanenbaum, *Structured computer organization*, 2nd edition, Prentice-Hall, London, 1984.

[49] J. E. Whitesitt, *Boolean algebra and its applications*, Addison-Wesley, Mass., 1961.

Chapters 8 and 10. The duality for distributive lattices is surveyed in [51] and [53]. The historical comments made there are supplemented by the introduction to [25] which puts Marshall Stone's pioneering contributions in perspective. A discussion of Stone's original, purely topological, representation of distributive lattices (equivalent to the order-topological theory we give in Chapter 10) can be found in [18] and [24].

[50] G. Birkhoff, On the combination of subalgebras, *Proc. Camb. Phil. Soc.* **29** (1933), 441–464. [Theorems 8.17 and 10.3.]

[51] B. A. Davey and D. Duffus, Exponentiation and duality, in [28], pp. 43–95.

[52] L. Nachbin, *Topology and order*, Krieger Publishing Co., New York, 1976.

[53] H. A. Priestley, Ordered sets and duality for distributive lattices, in [27], pp. 39–60.

[54] M. H. Stone, The theory of representations for Boolean algebras, *Trans. Amer. Math. Soc.* **40** (1936), 37–111. [Theorem 10.8.]

Chapter 11. Concept analysis is a new and fast-growing field. The topic was introduced by R. Wille in [28], pp. 445–470. It has been followed by a large number of papers, mostly written by R. Wille and other members of his Darmstadt group. The papers listed below provide an entry into the extensive literature, much of which is in German and appears in conference proceedings.

[55] R. Wille, Liniendiagramme hierarchischer Begriffssysteme, in *Anwendungen der Klassifikation: Datenanalyse und numerische Klassifikation*, H. H. Block (ed.), INDEKS Verlag, Frankfurt, 1984, pp. 32–51. (English translation: Line diagrams of hierarchical concept systems, *International Classification* **11** (1984), 77–86.)

[56] R. Wille, Lattices in data analysis: how to draw them with a computer, in [30], pp. 33–58.

[57] R. Wille, Bedeutungen von Begriffsverbänden, in *Beiträge zur Begriffsanalyse*, ed. by B. Ganter, R. Wille and K. E. Wolff, B. I. Wissenschaftsverlag, Mannheim, Wein, Zürich, 1987, pp. 161–211.

Notation Index

The symbol := denotes 'equals by definition'. The end of a proof is marked by ∎. The beginning of an argument by contradiction is flagged by 'suppose [by way of contradiction]' and the lightning symbol, $\lightning$, signals that the required contradiction has been reached. See for example the proof of Lemma 1.14.

Index

Page numbers given in boldface refer to definitions and those in italic to exercises, with the latter overriding the former where definitions are given in exercises.